iPHONE 12 PRO MAX COMPLETE USER'S GUIDE AT A GLANCE

Fast and Easy Way to Master the iPhone 12 Pro Max with Hidden Tips and Tricks

DAVID

GREAT

Copyright

All rights reserved. No part of this publication **iPhone 12 Pro Max Complete User's Guide At A Glance** may not be reproduced, stored in a retrieval system or transmitted in any form or by any means, electronic, mechanical, photocopying, recording, scanning without permission in writing by the author.

Printed in the United States of America
© 2021 by David Great

Contents

Copyright..i
Why This Guide?... vviii

Chapter 1

INTRODUCTION..1
Features 3
Build 3
Body Dimensions 3
Display..4
Resolution...4
Protection..4
Battery 4
Internal Storage 5
Main Camera Quad 5
Selfie Camera 5
Comms..5
NFC 6

Chapter 2

BONUS TIPS 7
Comparison With iPhone 12, iPhone 11 And iPhone 11 Pro Max 8

The Camera 10

The Back Camera 11

MagSafe 12

Setting Up The iPhone 12 Pro Max 17

5G Speedtest 23

Chapter 3

UNBOXING iPHONE 12 PRO MAX 28

What's in the box? 29

Chapter 4 33

FIRST 12 THINGS TO DO ON YOUR iPHONE 12 PRO & iPHONE 12 PRO MAX 33

Make Sure Your Battery Capacity Is 100% 34

Turn On Optimized Battery Charging Option 35

Enable Raise To Wake 36

Minimize Auto-Lock Time To Extend Battery 38

Customize Your iPhone With In-Built Widgets And Custom Widgets From Third-Party Widgets Makers 41

Enable The Bold Text Option 47

Turn Off Auto Brightness 48

Secure Your Phone Using Face ID 50

The Attention Features In The Face ID & Passcode Settings Menu 52

The Allow Access When Locked Features In The Face ID & Passcode Settings Menu 53

Disable Vibration To Extend Battery Life 55

Enable Picture In Picture (PIP) View........................56

Restrict Access To Apps And Other Areas On Your iPhone 61

Customize Your Control Center................................67

Chapter 5

25+ TIPS, TRICKS AND HIDDEN FEATURES FOR iPHONE 12 PRO AND iPHONE 12 PRO MAX 71

The Basics Of Navigation 72

How To Get A Customized Home Button 73

How To Back Up Your iPhone To iCloud..................77

How To Get Additional Cloud Storage......................78

How To Use The Dark Mode Feature In The Brightness Slider 80

New Wallpaper Features And How It Increases Battery Life..82

How To Customize Your Notifications Per Application 85

How To Take A Screenshot On Your iPhone And How To Use The Screenshot Customization Screen 87

How To Record Your Entire Screen 90

How To Modify The Pixels And Frames Of The Camera In Video Mode Without Leaving The Camera App In Seconds 96

How To Noties If An Application Is Using Your Device's Microphone Or Cameras In Any Form 98

How To Access The Different Formats In The Camera Settings 100

How To Enable The High Dynamic Range (HDR) And Dolby Vision 100

How To Access The App Library 102

How to Use Reachability feature 104

Smart HDR Three 106

How To Use Back Tap 106

How To Hide Page 107

Chapter 6

BEST ACCESSORIES FOR YOUR iPHONE 12 PRO & iPHONE 12 PRO MAX 109

MagSafe Accessories 110

Other Accessories For The iPhone 12 Series 110

Ailum Glass Screen Guard 112

Nomad Base Station Pro 113

Charging Brick 114

Manfrotto Clamps And Tripods 115

Chapter 7

WHICH APPLE WATCH IS RIGHT FOR YOU? IS IT THE SERIES 3, OR THE SE OR THE SERIES 6? 116

Apple Watch Series 3 118

Apple Watch Se.. 118

Apple Watch Series 6 119

Displays For All Three Watchs 121

Chapter 8

SIRI... 123

Siri 124

Siri voice... 125

Siri responses... 125

Always Show Siri Captions...................................... 126

Always Show Speech 127

Create Your Own Contact With Siri 127

Chapter 9

TIPS AND TRICK OF THE CAMERA OF THE iPHONE 12 PRO MAX 129

How To Enable High Efficiency 130

How To Enable Record Video 131

How To Enable Lock Camera 132

How To Turn On Record Slo-Mo 133

How To Use Record Stereo Sound 134

How To Use Grid 134

Mirror Front Camera And View Outside The Frame
.... 134

Three Ways To Clear And Free iPhone Ram Memory
On iOS 14 135

Step 1 135

Step two 136

Step Three 136

About the Author 138

Why This Guide?

If you need a regular iPhone that does everything that the top of the line iPhone Pro does for four times less the price, then the 6.7-inch iPhone 12 Pro max is your best bet. However, if You need a comprehensive guide to walk you through the essential settings, configurations and numerous handy tips, tricks, hidden features and various techniques of the iPhone 12 Pro max Device, then this guide is for you. It provides an insight into the basic functions of the iPhone 12 Pro max such as FaceID, live radio, Airdrop, ScreenTime, etc to advanced functions such as creating Advanced Siri commands, advanced gestures, setting up advanced security and techniques to master various advanced settings to safeguard your device and increase productivity.

This book also gives you insight to several useful accessories for the iPhone you should purchase and seamless techniques to connect your iPhone 12 Pro max to several hardware devices such as external monitors, airpod and several advanced hacks that would push your iPhone 12 Pro for maximum performance.

.

Chapter 1

INTRODUCTION

The iPhone 12 Pro and the iPhone 12 Pro Max are unique and clerical new design beast from Apple, with a new design, a reengineer chip and a 5G features. The iPhone 12 Pro and 12 Pro Max vertical band are precision machined from stainless steel. The ceramic shield on the front is also a durable and robust glass than its predecessor. Its tighter border allows super Retina XDR displays.

The 5G network is faster and speeds set to change everything about the smartphone device. The processor, Hardware and software have been optimized for the 5G experience. Its A14 Bionic has a smaller five-nanometer transistor that allows the powerful chips in a smartphone generation ahead. It is built with a new Camera system which will makes the photographers iPhone.

The larger iPhone 12 Pro Max takes the innovations even bigger and further, the bigger sensor increases its details captures and now helps improve low light performance in dark places or night by 88percent. The customs designed LiDar scanner adds advance feature tech in your pocket; you can also take a night mode portrait with it. I bet you the iPhone 12 Pro Max smashes the highest quality video benchmark in a smartphone and its water resistance. It is a device that can capture playback and edit 10-bit HDR footage with Dolby Vission. The device has an intelligent system that magnet connect accessories.

Features

Build: Glass front (Gorilla Glass), glass back (Gorilla Glass), stainless steel frame

SIM: Single SIM (Nano-SIM and/or eSIM) or Dual SIM (Nano-SIM, dual stand-by) -for China. IP68 dust/water resistant (up to 6m for 30 mins)

Body Dimensions: 160.8 x 78.1 x 7.4 mm (6.33 x 3.07 x 0.29 in)

Weight: 228g (8.030 ounce), height 6.7", **thickness** 7.4mm

Size: 6.7 inches, 109.8 cm2 (~87.4% screen-to-body ratio)

Display: Type Super Retina XDR OLED, HDR10, 800 nits (typ), 1200 nits (peak), Display Contrast ratio: Infinite (nominal). Chipset Apple A14 Bionic (5 nm)

CPU: Hexa-core (2x3.1 GHz Firestorm + 4x1.8 GHz Icestorm)

GPU: Apple GPU (4-core graphics)

Resolution: 1284 x 2778 pixels, 19.5:9 ratio (~458 ppi density)

Protection: Scratch-resistant ceramic glass, oleophobic coating Sensors, Face ID, accelerometer, gyro, proximity, compass, barometer

Siri natural language commands and dictation

Stand-by: Up to 20h (multimedia), Dolby Vision

Wide color gamut, True-tone and have Apple A14 Bionic

NETWORK Technology 5G

Battery: Type Li-Ion 3687mAh, non removable (14.13 Wh), Charging Fast charging 20W, 50% in 30 min (advertised) USB Power Delivery 2.0 Qi magnetic fast wireless charging 15W

Platform: OS, iOS 14.1, upgradable to iOS 14.2

Memory: Card slot No

Internal Storage: 128GB 6GB RAM, 256GB 6GB RAM, 512GB 6GB RAM, 12MP 2160p 6GB RAM

Main Camera Quad: 12MP, f/1.6, 26mm (wide), 1.7µm, dual pixel PDAF, sensor-shift stabilization (IBIS), 12 MP, f/2.2, 65mm (telephoto), 1/3.4", 1.0µm, PDAF, OIS, 2.5x optical zoom, 12MP, f/2.4, 120°, 13mm (ultrawide), 1/3.6", TOF 3D LiDAR scanner (depth)

Features: Dual-LED dual-tone flash, HDR photo/panorama

Video: 4K@24/30/60fps, 1080p@30/60/120/240fps, 10-bit HDR, Dolby Vision HDR (up to 60fps), stereo sound rec.

Selfie Camera: Dual 12 MP, f/2.2, 23mm (wide), 1/3.6" SL 3D, (depth/biometrics sensor) Video 4K@24/30/60fps, 1080p@30/60/120fps, gyro-EIS, Music play Up to 80h

Sound: Loudspeaker Yes, with stereo speakers 3.5mm, jack No

Comms: WLAN Wi-Fi 802.11 a/b/g/n/ac/6, dual-band, hotspot, Bluetooth, 5.0, A2DP, LE

GPS : Yes, with A-GPS, GLONASS, GALILEO, QZSS

NFC yes, Radio No, USB Lightning, USB 2.0

MISC Colors Silver, Graphite, Gold, Pacific Blue

Models A2411, A2342, A2410, A2412

SAR EU: 0.99 W/kg (head) 0.99 W/kg (body)

Price $ 1,099.00 / € 1,217.50 / £ 1,099.00 / ₹ 126,900

Chapter 2

BONUS TIPS

Comparison With iPhone 12, iPhone 11 And iPhone 11 Pro Max

The blue color of the iPhone 12 is very vibrant compared to the iPhone 12 Pro's Pacific blue. They are the same size, fit the same cases. Everything is the same, except for the camera and a few other specs.

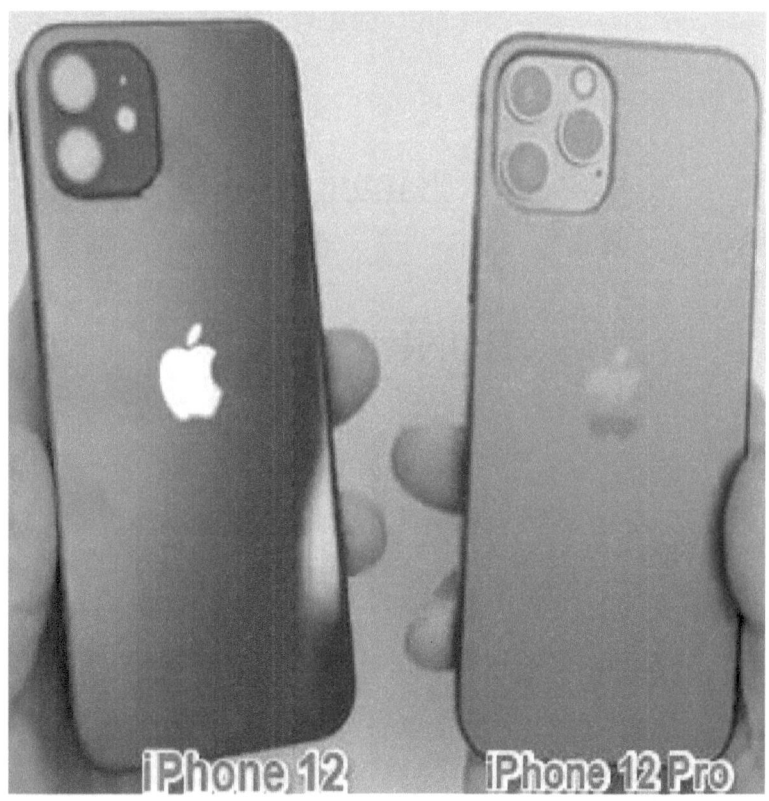

Compared with the iPhone 11 pro, they are about the same weight, the top of the devices is very similar, and they are very close as far as width goes. It just feels a little bit thinner with the iPhone 12 Pro.

Finally, compared with iPhone 11 Pro Max, there is a difference in size. There is a massive difference in size, the width is narrower, and the height is much less.

The Camera

On the back, one of the advantages of the iPhone 12 Pro is the LiDAR (Light detection and ranging) sensor; it helps with augmented reality, even autofocus in dark situations, when you are at night, and it has three (3) different cameras.

The 12 MexaPixel (MP), Ultrawide with F/2.4 aperture, with 120 degrees field of view and the 12MP Wide with F/1.6 aperture, and that's very similar to what we have with the iPhone 12, which lets in a little bit more light. Also the 12MP Telephoto with F/2.0 aperture, but it has 2x optical zoom and then 4x sort of zoom range, probably punching into the sensor the actual phone itself or the camera. It also has the night mode with portrait mode, which is vital.

Then, of course, on the back, all the cameras record in 4K 60. They have Dolby vision, HDR as well, and Apple Pro raw is coming later this year. They record in 10 bit HDR, which is impressive.

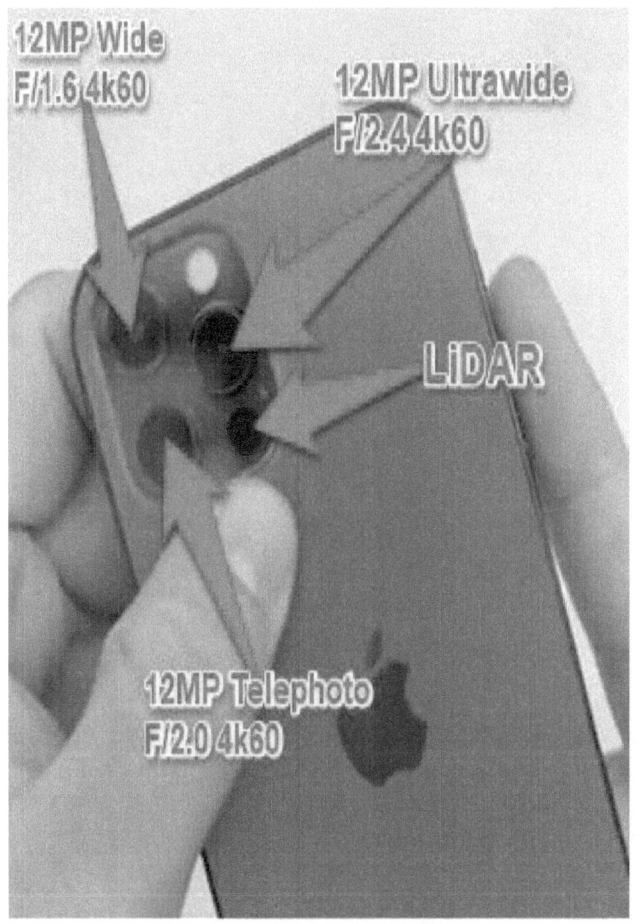

The Back Camera

The forward-facing camera is very similar to the iPhone 12. It's a 12MP true depth camera with an F/2.2 aperture. Again, it can record in 4K 60.

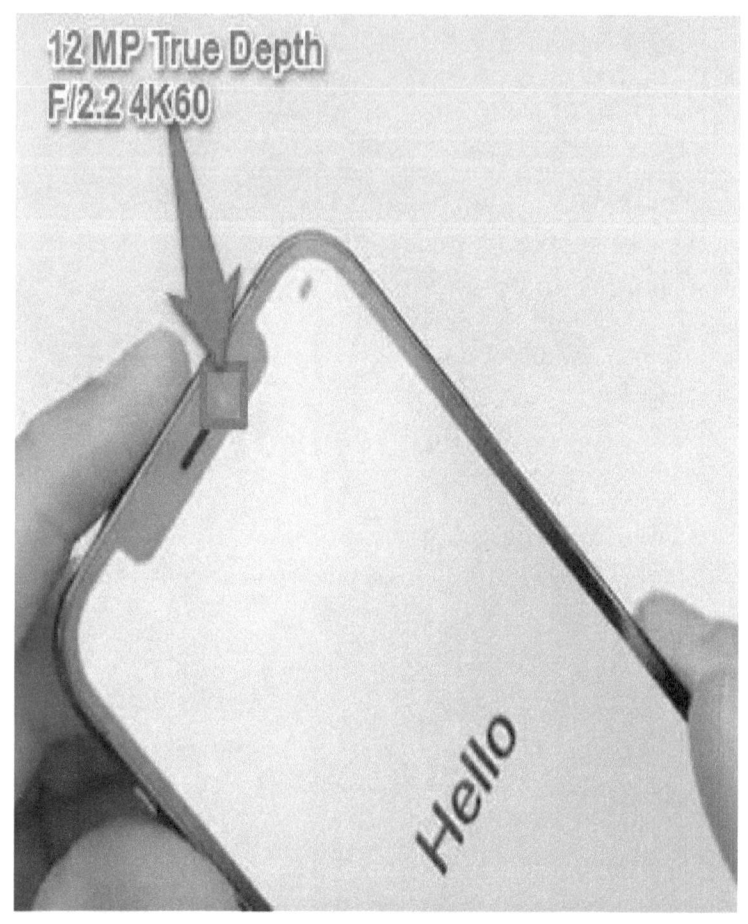

MagSafe

Now on the back of the phone, we have the MagSafe, it's its new way of charging wirelessly with the magnet paper. There are the speaker and microphone with the magnet paper and then the ring in the middle as you move the magnet paper around. The ring in the middle is for MagSafe.

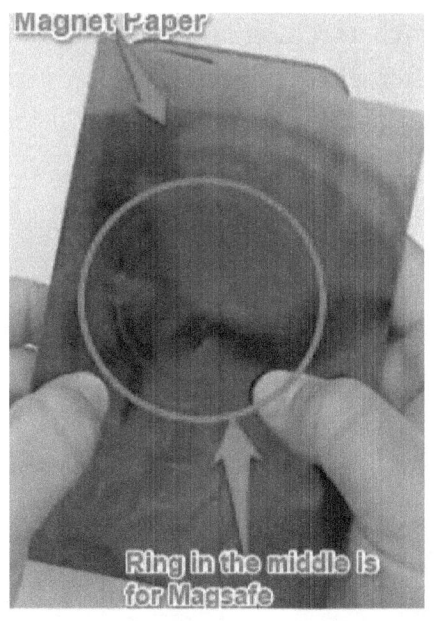

The MagSafe allows you to fast charge up to 15watts.

When you attach the MagSafe to charge the phone, you will hear a different sound, and it adheres pretty tightly to the back of the phone. You can jiggle it around, and it is not going to drop although, it is not recommended that you throw it; it might drop that way.

Now, MagSafe is not only for charging; it does some interesting things with cases. So, Apple is using a magnetometer and NFC to recognize the devices that are connected to it. The plum case will fit the iPhone 12 or iPhone 12 Pro.

Lock the display and then put on the Plum case. You will discover that the MagSafe shows up as plum when it connects to the case. It is using that NFC in the back to do that.

If you take the case off and turn off the display, put on the deep navy case. The MagSafe connects, and it shows you the navy color, so it knows that it is connected to this case.

Of course, you can charge through it, and the magnets hold on tight, especially with the case, because the case is doubling the magnet's power, so it is really on there. You can easily pop off the MagSafe to remove it.

These cases are fantastic because they protect the bottom of the device this time too.

Now we have MagSafe to quick charge at 15Watts and 7.5Watts on a regular charger. So, if you want to use a regular charger, a wireless charger, you still can; of course, use the lightning adapter as well, but you will need another charger of some sort. You will have to pick one up.

Now when you connect MagSafe, there is this nice animation. Connect it to the back of the device, flip the device over, and you will see it shows the actual charge state.

It also has a nice sound you will hear when the device shows the actual charge state to be aware the device is charging. It is a nice animation to go along with it.

Setting Up The iPhone 12 Pro Max

To set the iPhone 12 Pro up,
- Press and hold the power button until the Apple logo appears.
- Enter the passcode of your other iPhone. Once you put in the password, it will start setting it up, and on the old iPhone, it will say "Finish on the New Phone," and you can set the old iPhone aside. Now, it may take a few minutes to activate the device. You can put in your sim card at any time, but what this does is bring over all the information, such as your WI-FI network and other settings as well, so it's nice that the iPhone has that, just an easy way to move that over.

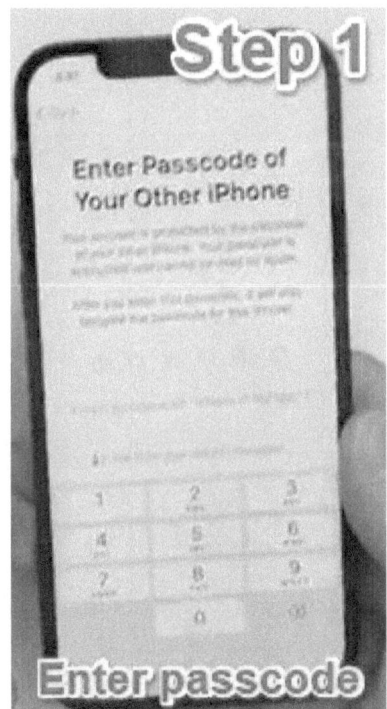

- Next, it will ask you to set up "Face ID," tap on the option listed as continue, and on the next page that pops up, click on "Get Started."

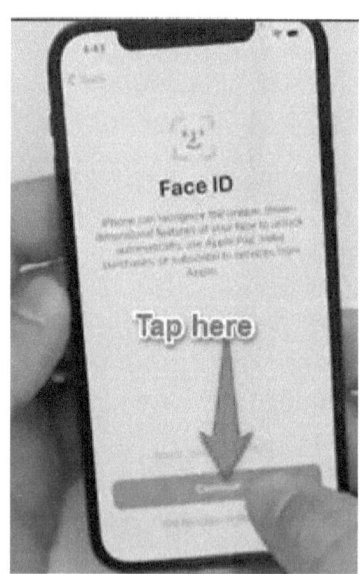

- You will see that a camera appears tha grabs onto your face; move your face around slowly to complete the circle and tap on the option listed as "Continue," and move your face around again. Once your face is fully scanned the second time, it will state, "Scan complete." So, now your face is registered for Face ID.
- The device will take you to a new page with a "good mark" and a message reading "Face ID is now set up." Click on the button listed as "Continue."

- Then the device says, "Setting up your Apple ID."
- Next, you have to agree to the terms and conditions.
- At the next point, on the page listed as "Apps & Data," the device will list options asking you if you want to restore your apps and data from an iCloud backup or transfer apps and data. For now, click on

the 'Dont transfer Apps & Data" option so that we can get to the main home screen.
- Next, on the page listed as "Settings from your Other iPhone," the device will have a message on the page asking you if you want to bring over settings from your other iPhone; click on the option listed as "Continue."

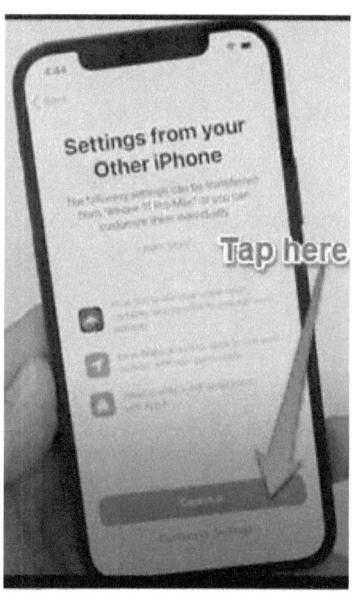

- Next, a page with the heading "Keep Your Phone Up to Date," will appear with a message on the page, asking if you want to keep your iPhone up to date. There is no option to opt-out of this; the only option listed out is "Continue;" you will have to do it a little bit later. Before, there used to be an option for that, but it isn't available on this device.
- Next up is "Apple Pay," click on "Set it up later."
- Then, "Improve Siri & Dictation," appears so you can "Share Audio Recordings" or tap "Not Now."

- Next up is "App Analytics," you can tap on "Share with app Developers" to share the apps' analytics or tap "Not Now."
- The next step is "Display Zoom," hit continue, and the device says, "Welcome to iphone," and at the bottom of the screen, it says, "Swipe up to continue."

- When you swipe up, it takes you to the device's home screen.

- If there is no sim card installed, a message will pop up immediately after the Home Screen launches, saying: No sim card installed.
- Another message will pop up, asking if you will "Use This iPhone When Sharing Your Location?" tap on "Use."
- Next, tap on Settings on the home screen, and it will take you into the Device Settings Menu; and from there, you can see the device is now set up.

To take a look at the actual software version of the device pre-installed, follow these steps:
- On the Settings menu, tap on "General" and then tap on "About."
- You can see the "Software version" pre-installed at the top of the About page is iOS 14.1 Build 18A8395, which is pretty expected.

5G Speedtest

For the 5G Speedtest, follow these steps:
- Download Speedtest if you don't have it on your device.
- After installing it, launch the application to set it up.
- Ensure your sim card is installed.
- If you are using Wi-Fi, disconnect it.
- On Speedtest, you can see that the device is on 5G, and this device has two bars on the 5G radar.

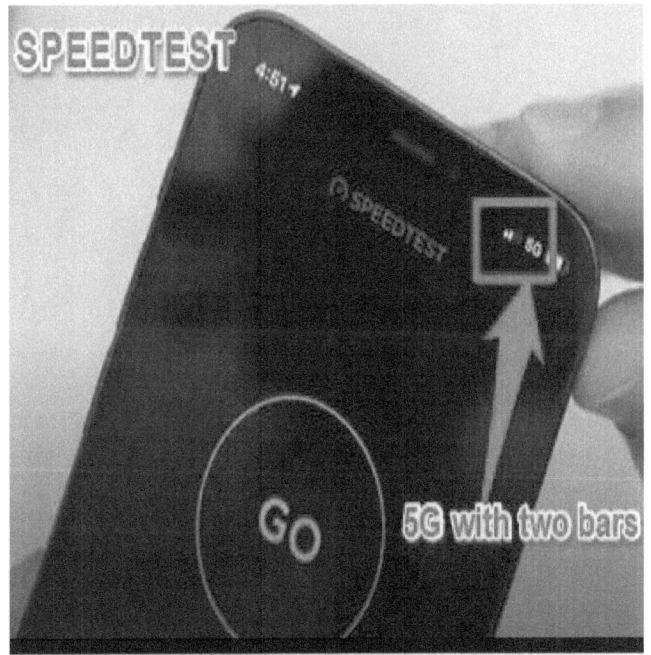

- Tap "Go" on the Speedtest, and it will start connecting. This connection will ramp up to 5G or back to 4G Lte depending on if it is at an advantage or not.

- This device hit 135megabits per second download, and give it just a moment, and we will see what it does for upload. It is getting to about 10-11 megabits per second. But of course, you can do much more with that if you have got a better signal, and if you have millimeter-wave, it will be better.
- On this device, after the Speedtest, we got 114megabits download by 11.6megabits upload. If you have millimeter-wave, that will come into play, and you can get gigabit speed in the Speedtest. But in the United States, it is mostly Verizon that has that.

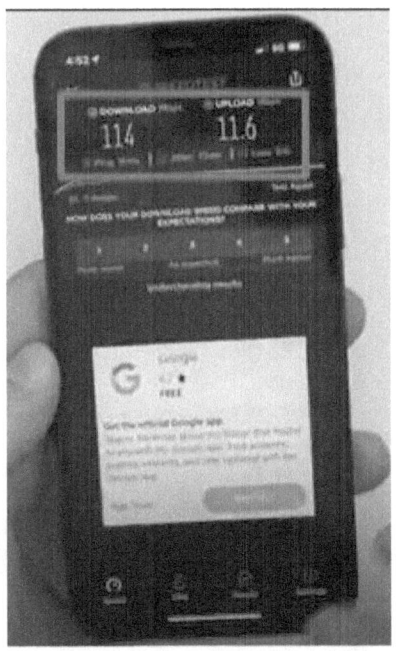

Now to customize the 5G Settings, follow these steps:

- Go into the device Settings, go into the option listed as "Cellular," and tap on "Cellular Data Options."

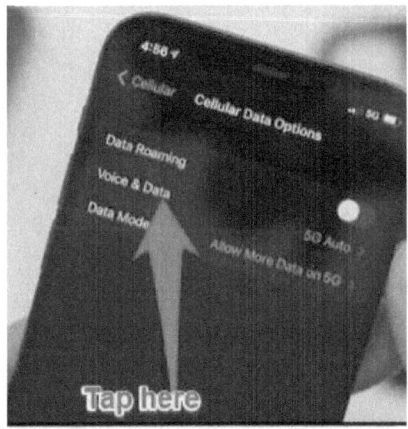

- Now within the "Cellular Data Options," when you tap on "Voice & Data," you have "5G Auto" set as default .

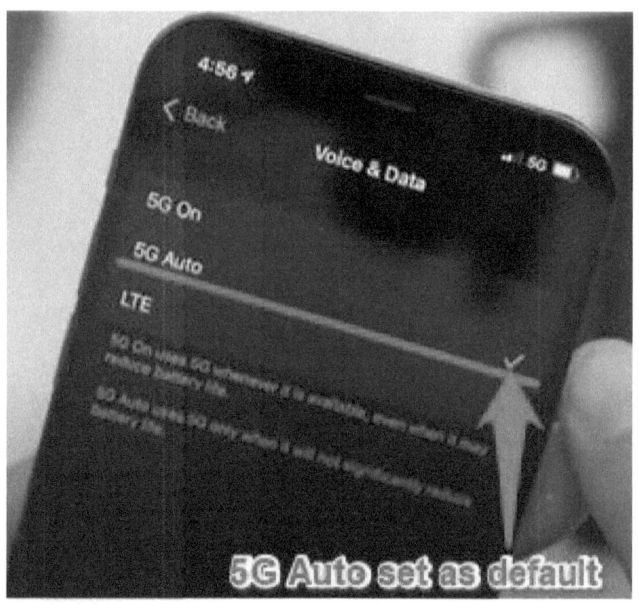

- Now the Setting will switch based on the need to try and save you the most power, but you do have the option to turn "5G On" or switch back to "LTE" full time.
- Also under,"Data Mode" in the "Cellular Data Options," you have "Allow More Data on 5G," "Standard," or "Low Data Mode." Those are the options within the phone as well. That may or may not help the battery, depending on what it is doing. Apple should be managing that by itself.

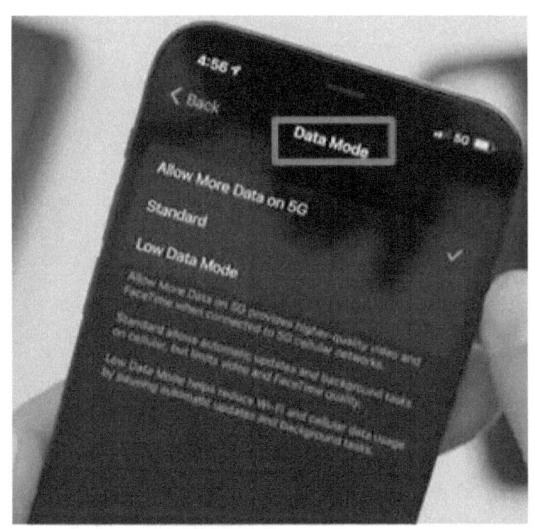

Chapter 3

UNBOXING iPHONE 12 PRO MAX

The iPhone 12 Pro comes in four different colors: Graphite, which replaces Space Grey, Silver, Gold, and the all-new Pacific Blue.

It also comes in at Nine Hundred and Ninety-Nine Dollars ($999) for 128 gigabytes (GB) of storage, One Thousand and Ninety-Nine Dollars ($1099) for 256GB of storage, and One Thousand, Two Hundred and Ninety-Nine Dollars ($1299) for 512GB of storage.

What's in the box?

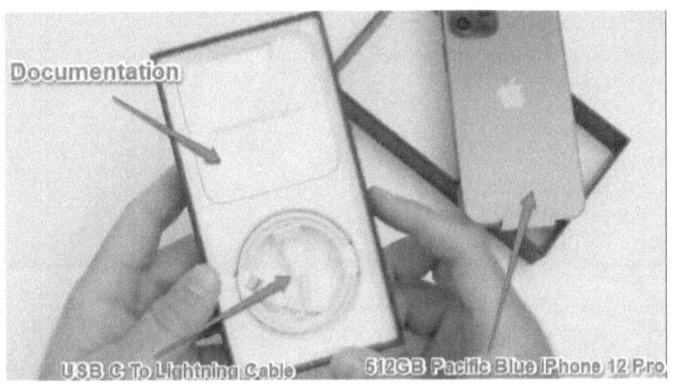

This discussion is based on a 512GB Pacific Blue iPhone 12 Pro.

The box has a colored logo, and it is very similar sized to the iPhone 12. It's about two-thirds the size of an iPhone x ebox or the current one because it doesn't include a charger.

Inside the box, have the 512GB Pacific Blue iPhone 12 Pro, a USB C to lightning cable, and a smaller pamphlet-sized envelope. Inside the envelope, we have the sim card injector. tool, a little information about the warranty on a small pamphlet, and a single white Apple sticker.

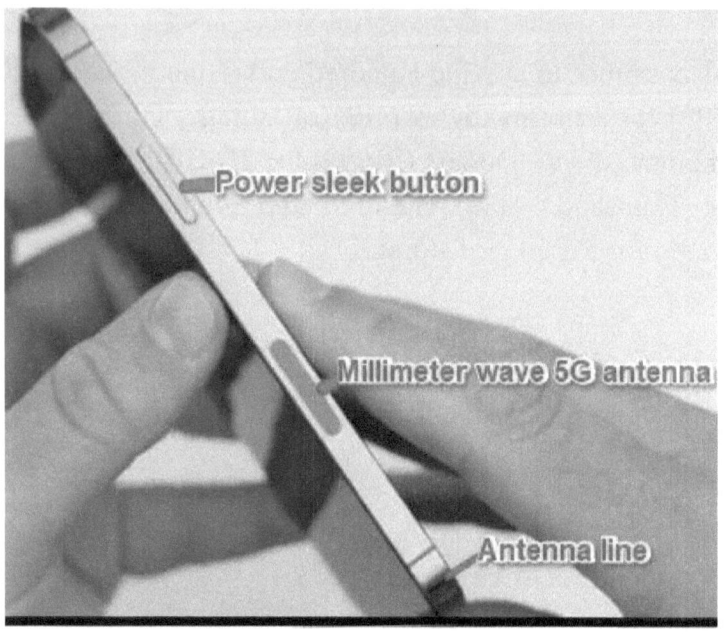

The Right View

The phone doesn't have any wrapper on it, other than the piece of paper on the front display. When you take that off, you discover it's the same display and exact size as the iPhone 12.

The phone weighs 6.66ounces or 189grams, so it's heavier than the iPhone because it has surgical stainless steel, according to Apple, around the outer edges.

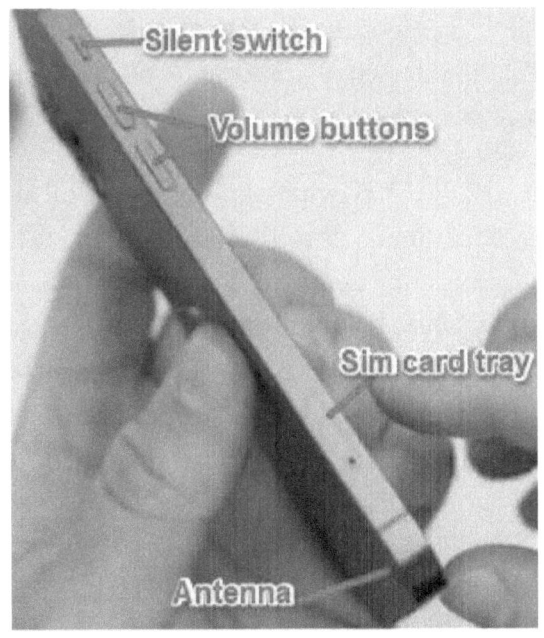

Bottom View

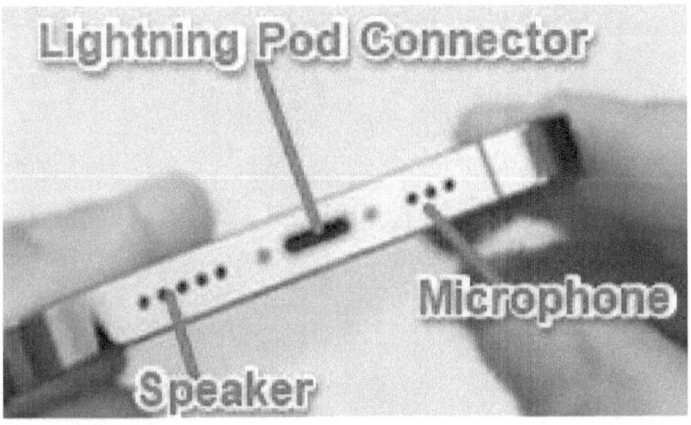

Just from touching it, you will discover that the outside rim is a fingerprint magnet, but the back is not.

The overall spec has the all-new Bionic A14 CPU inside, running at 2.99GHz (GigaHertz) and 6GB of RAM. It has 5G with millimeter-wave (mm-wave).

The iPhone 12 Pro has what should be an all-day battery, but 5g may affect that. It is IP 68 rated, but it has a better rating of a maximum depth of 6 meters up to 30 minutes, which is impressive. The battery is about the same as the iPhone 11 Pro, but it may get a worse battery. Some people say one to two hours worse depending on everyday use, so it depends on what you are doing with the device.

Chapter 4

FIRST 12 THINGS TO DO ON YOUR iPHONE 12 PRO & iPHONE 12 PRO MAX

We will discuss the first 12 things to do on your brand new and shiny iPhone 12 Pro and iPhone 12 Pro Max. You want to make sure that you tweak all these settings to get the best experience, security, and battery life out of your iPhone.

Make Sure Your Battery Capacity Is 100%

The very first thing you should do is:

- Go into your Settings
- Once you are in your Settings, tap on the Battery option.
- In the Battery Settings, tap on "Battery Health."
- In the Battery Health menu, take a look at the number beside the "Maximum Capacity" option.
- The Maximum Capacity should say one hundred percent (100%).
- When you buy a brand new phone, the "Maximum Capacity" of that phone should always be a hundred percent for the battery.

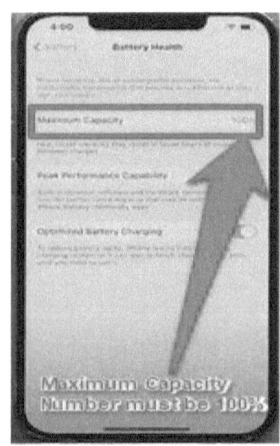

- If the Maximum Capacity number is any lower than a hundred percent, there is a problem with your battery, and that means you want to return that iPhone and get a replacement quickly.
- Again, even ninety-nine percent (99%) is not acceptable when you are supposed to be getting a brand new product.
- And, of course, over the years, the Maximum Capacity Number will go down a little bit but not by too much.
- From a brand new phone perspective, even ninety-nine percent (99%) means terrible.

Turn On Optimized Battery Charging Option

The number two thing you need to do is this:

- While you are still in the "Battery Health" options menu, make sure the "Optimized Battery charging" is enabled by clicking on the nub beside the option.

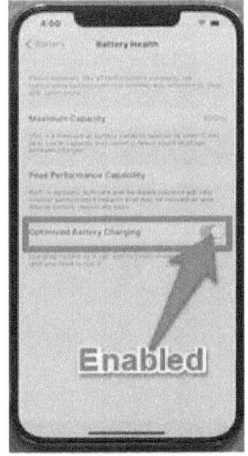

- When the "Optimized Battery Charging" is enabled, the iPhone optimizes the charging of your battery. So, once the battery charging hits eighty percent (80%), the iPhone will throttle the charging speed to ensure the charging itself does not damage the actual battery in the long term. So you have a pleasant and healthy battery for a long time.
- When the "Optimized Battery Charging" option is enabled, the overall charging from 0-100% takes a hit. So the charging speed slows down a little bit, but the life of the battery prolongs itself.
- Suppose you disable the "Optimized Battery Charging" option. In that case, the battery will charge a little faster, and it also takes a hit in overall battery life, overall longevity.

Enable Raise To Wake

One more important thing to set up is to enable the "Raise to Wake" option. To enable it, follow these steps:
- Go into Settings, then go into "Display."
- Ensure the "Raise to Wake" option is enabled by clicking on the nub beside it.

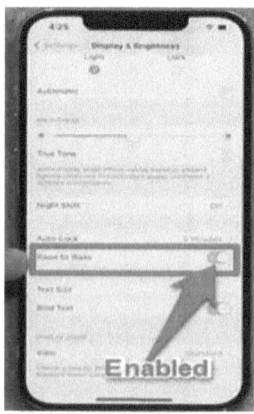

- When the "Raise to Wake" option is disabled, lock the phone screen.
- When you grab the phone and lift it, as the device is locked, nothing happens. If you want to glance at the device, nothing happens.

- When you have the "Raise to Wake" enabled, lock the phone screen.
- Now, when you grab the phone and want to glance at it, the phone wakes up and displays the lock screen, and you can check if you have any new text messages or any missed calls.

- And when you put the device right back, the device goes back off.
- The good thing about this option is that you can raise to wake, and it's going to show you what's happening on your device.
- If you are not satisfied with the option, you can disable the option.

Minimize Auto-Lock Time To Extend Battery

- Go to your Settings. Then go into "Display and Brightness."
- In the Display and Brightness Settings, ensure that "Auto Lock" is set to 30 seconds.
- Tap on the "Auto-Lock" option, and a list of options will open in a new menu.

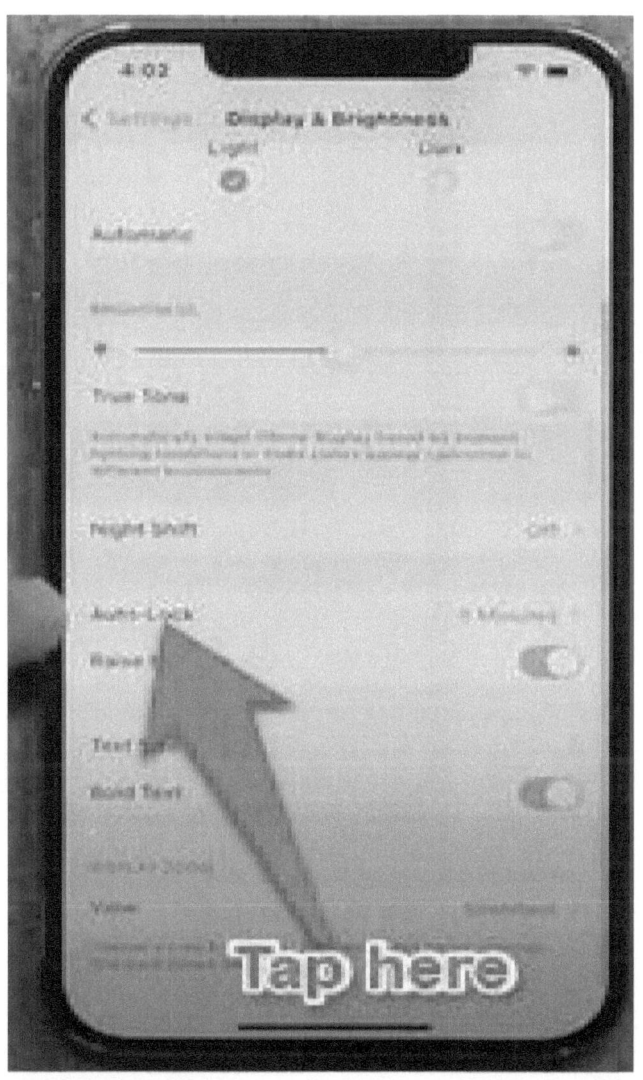

- In the Auto-Lock menu, tap on "30 seonds." This option is going to make sure you get the maximum battery life per charge.

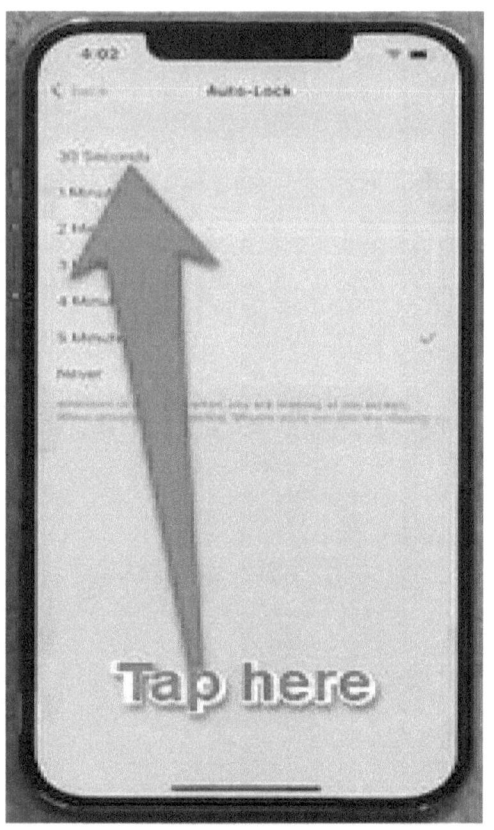

- When Auto-Lock is at 30 seconds, if nothing is happening on the device's screen, nobody is touching the screen, the device turns off automatically in 30 seconds.
- But if you set the Auto-Lock to "Four minutes," or "Five minutes," or "Three minutes," the device is just going to stay on for no reason, and it is only going to eat the battery life.
- As you know, the display is what consumes most of your battery, so keep the Auto-Lock at 30 seconds.

Customize Your iPhone With In-Built Widgets And Custom Widgets From Third-Party Widgets Makers

The next thing you want to do with iPhones with iOS 14 is to add widgets onto your device's screen, just like you have on the android phones.

To add widgets onto your device's screen, follow these steps:
- Swipe over on the device's screen, and you will discover that this device has a bunch of widgets. Now, it is very straight-forward to add widgets to the screen.

- Press and hold the device's screen, and the whole screen will wiggle.

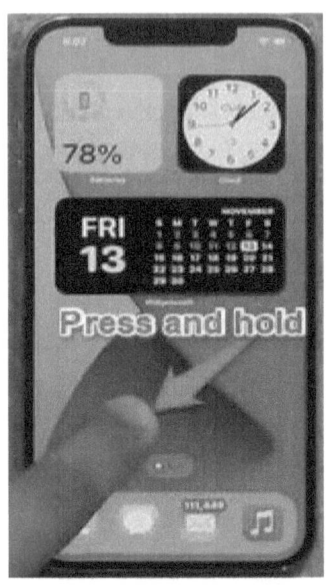

- Click on the "Plus" icon that appears on the top left corner of the screen.

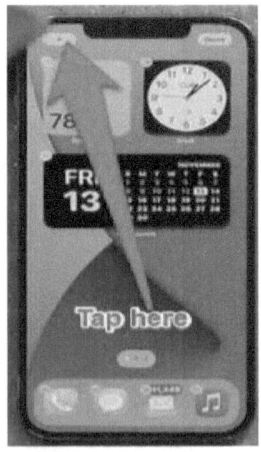

- A menu will pop up with many widgets you can choose from there for all kinds of applications.

- Most of these widgets in this menu are built-in widgets.
- For example, if you want a "Weather widget," click on the weather widget in the menu, and that will take you to a new Weather menu, from there; you can choose the various sizes available by swiping the screen.

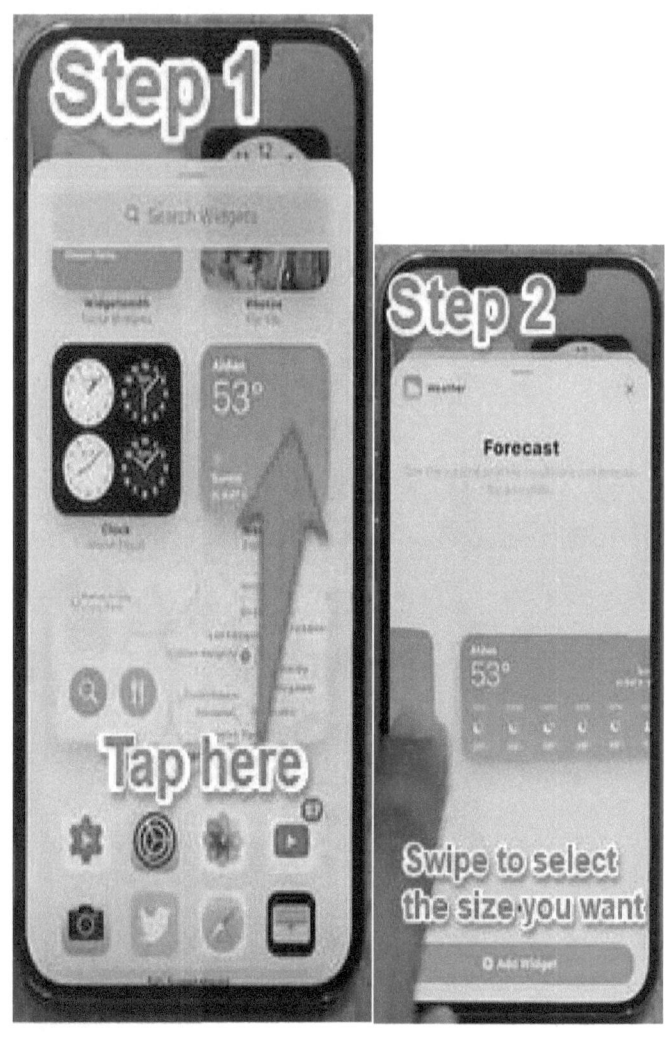

- Once you have settled on the size of the widget you want, you can click on the option listed as "Add Widget" at the bottom of the screen to add the new widget to the page on the screen showing the widgets.

- You can also add new widgets to your device by pressing and holding the widget, and it will take you back to the screen showing the widgets on the device where you can drag and drop the widget where you deem best.

Note: These in-built widgets on the device are not going to be enough for maximum customization. What you also want to do is to go for custom widgets from third-party widget makers. To do that, follow these steps:
- Go to the App Store and search for Widgets.

- When you search for Widgets on the App store, you will get access to all the applications that allow you to create additional super customized widgets. For example, the "Widgetsmith."

- So you can do all kinds of things with your widgets if you download third-party widget makers from the App Store. Don't just rely on the inbuilt widget; also, go for custom widgets.

Enable The Bold Text Option

The text on the device is very thin; to bolden it, and to be able to see better on a larger screen, follow these steps:
- Go to Settings and go into Display and Brightness.
- Enable the option listed as "Bold Text" by clicking on the nub beside it. That will give you an excellent black pronounced text, which is will be much easier to see.

Turn Off Auto Brightness

There is an important feature that you need to deactivate or activate based on your needs. By default, the Auto-Brightness is activated. To deactivate it, follow these steps:

- Go to Settings and go into "Accessibility."
- Once you are in "Accessibility," go to "Display & Text Size."
- At the bottom of the "Display & Text Size" Settings Menu, the option "Auto Brightness" is hidden there, and it is enabled by default.

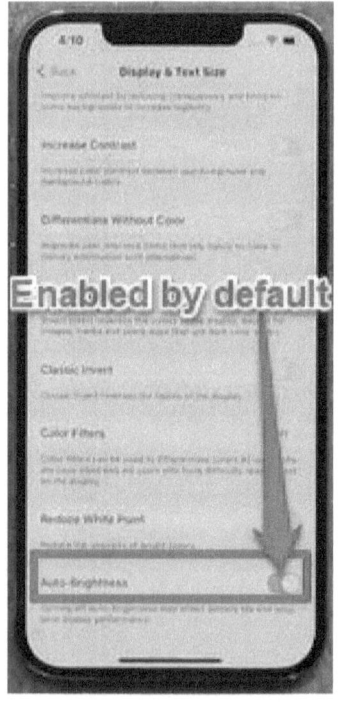

- To disable it, click on the nub beside it. When you disable the Auto-Brightness option, you can control the brightness of the device manually.

- When you go into the Control Center, by swiping the screen down from the top right corner of the display, there is a rectangular bar with a brightness icon embedded in it besides the volume bar on the screen's left side.

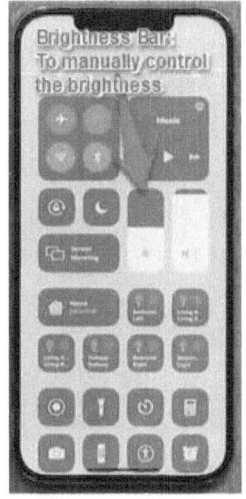

- You can increase or decrease the brightness by dragging the bar upwards or downwards as you desire.

Note: When the Auto-Brightness option is enabled, the device is just going to adjust the brightness based on the ambient lighting, without your input.

With the short message written below the Auto-Brightness option, you may discover the benefits of enabling Auto-Brightness as default. But manual control is preferable if you want to manage your battery life.

If you are looking at your phone in bed at night time, make sure you lower the brightness. That is going to keep your eye safe, and also, it is going to get you better battery life.

Secure Your Phone Using Face ID

The next thing you should do is to set up certain settings in your Face ID. To do that, follow these steps:

- Go to Settings and go into "Face ID & Passcode."
- Once you tap on the Face ID & Passcode Settings option, a new page will appear, requesting for your Passcode.
- Put in your passcode and it will take you directly to the "Face ID & Passcode" Settings Page.

The Use Face ID For Feature In The Face ID & Passcode Settings Menu

- When you are in the "Face ID & Passcode Settings" Page, you want to ensure that you enable all the options under the "Use Face ID For, Feature" namely, "iPhone Unlock," "iTunes & App Store," "Apple Pay" and "Password Autofill."

- That means when all the options under the "Use Face ID For" Features are enabled, you can unlock the iPhone when you use Face ID, you can use your Face ID to make purchases in iTunes and App Store, also for Apple Pay and Password Autofill, which is going to make things very convenient.

The Attention Features In The Face ID & Passcode Settings Menu

- Another important thing which is very important for security is, you want to make sure that the two "Attention" features are enabled, namely, "Require Attention for Face ID" and "Attention Aware Features."

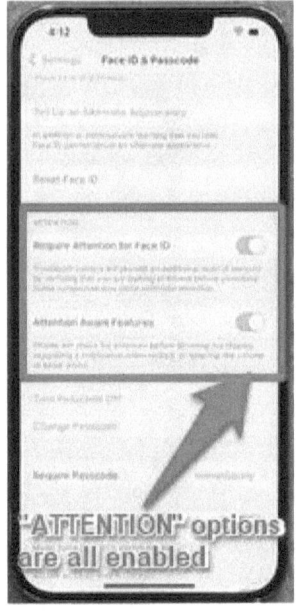

- These "Attention" Features when enabled, ensure thst when you are sleeping, somebody cannot bring the phone to your face and unlock the phone with your face.
- These "Attention Features" are going to ensure that your eyes are opened and looking into the camera to make sure Face ID unlocks the phone. This is definitely a must have for security, as it will ensure, no one can unlock your device using Face ID while you are sleeping and go through it.

The Allow Access When Locked Features In The Face ID & Passcode Settings Menu

- When you go down the "Face ID & Passcode" Settings Menu, you will come across the "Allow Access When Locked" Features. When all the options under these features are enabled, you can access all the options listed in this feature even if the device is locked, i.e., Today View, Notification Center, Control Center, and all the other options listed in the feature.

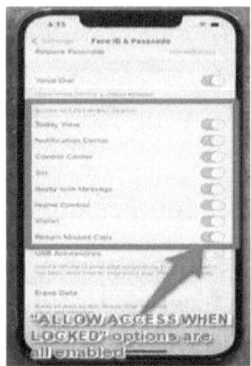

"ALLOW ACCESS WHEN LOCKED" options are all enabled

- For example, when you lock the device, double-tap the screen to wake it up. It will display the lock screen. From the Lockscreen, you can access the control center, and you can make modifications to all the options displayed on it, i.e., you can turn the lights on and off in the house right from the phone, even though the phone is supposed to be locked.

- People can also access the camera and take a photo by tapping on the camera at the bottom right corner of the lock screen to launch it.
- So what you should do is, go back into your phone, go back into the Face ID & Passcode Settings Menu and disable all the options listed under the "Allow Access When Locked." Click on the nub beside each option, so nobody, including you, can access these options when the phone is locked.

- People can even reply to your messages if you do not have these "Allow Access When Locked" options disabled.
- Now, when you lock the phone and wake it up, you can't access the control center. You cannot access anything; all you have to do is log in to get full access to all the functions.

Disable Vibration To Extend Battery Life

- Go into Settings, and go into "Sounds and Haptics."
- In the Sounds and Haptics Settings Menu, ensure you disable the "Vibrate on Ring" option by tapping on the nub beside it.
- By disabling the "Vibrate on Ring" option, you are saving battery because the vibrate motor in the phone uses a lot of battery. So, if you get a lot of calls, this option will save your battery life.
- Ensure you enable the "Vibrate on Silent" option so that when the device is on silent, it will vibrate so

that you can feel the device by the vibration or the vibrating sound.

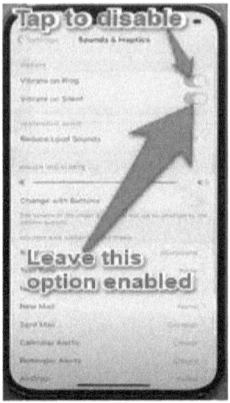

Enable Picture In Picture (PIP) View

- Go to Settings, tap on the "Search" at the top of the Settings menu and search for PIP and it will bring out the PIP option.

- Even if you do not search for PIP, you can find it in the Settings Menu, under General.
- In the General Settings menu, tap on the "Picture in Picture" option, and a new menu will pop up.

- Ensure that the "Start PIP Automatically" option is enabled in the Picture in Picture Settings Menu.

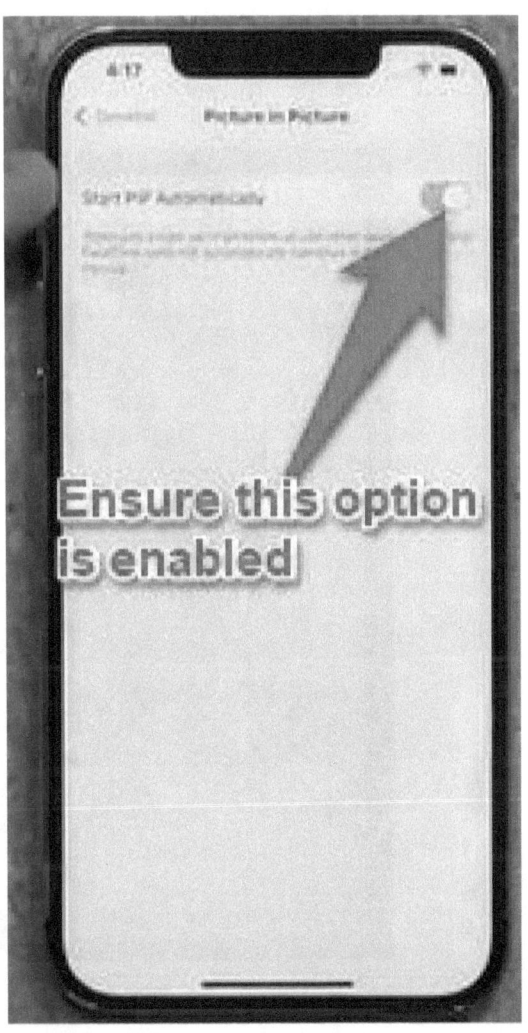

- For example, launch an application Netflix and pull up a movie and start playing it; as the movie is playing, you can do something else on the same screen if PIP is enable

- what you can do when PIP is enabled is when you pull the movie playing up, the movie continues playing on a little window on the screen, then you can do check massages and open another app on your phone as the movie is playing.

- For example, when you launch another application, i.e., Calculator, and you want to do a quick calculation as a thought occurred to you, you can do the calculation as you are watching the movie, which is excellent.

- You can also resize the window the movie is playing in (make it small, middle, or large) by pinching in or out on the device's window. Pinching in or out of the device window with your fingers' movement increases or decreases its size.
- You can also move the movie window to the device screen's side if the movie is not that important to watch. Let's say you want to access the full application you launched (in this case, the Calculator). Without pausing the movie, you can pull the window to the side of the screen, and the navigation button will be visible at the side of the
- screen you pulled the window to, to indicate the window is still there.

- Then you can do whatever you want to do, in whatever application you are in, and when you are done, tap on the navigation button to pull out the movie and go back to it.
- When you are done with the movie, you can exit it by tapping on the window to reveal the two icons at the top of the window. Then you can tap on the "x" icon at the top left corner of the window, or you can go back to the actual full screen by tapping on the fullscreen icon at the top right

corner of the window, it is all going to be up to you.

Restrict Access To Apps And Other Areas On Your iPhone

- The next important thing is the ability to restrict things on your iPhone. So, when you give your iPhone to your friends or family members, they can only do so much on it without causing damage or looking into things you don't want them to look in. You have to set this restriction from the very beginning. To restrict apps and specific areas on your iPhone, follow these steps:
- Click on the Settings, tap into "Screen Time."

- In "Screen Time," scroll down to the bottom, and the first thing you have to do is "Set Up Screen Time Passcode" by tapping on the option and setting up the passcode of your choice.

- **Note:** Now, this passcode is not the passcode you use to log into your phone; this is a unique passcode just for this Content & Privacy Restriction.
- Once you have done that, tap on the "Content and Privacy Restrictions" options.

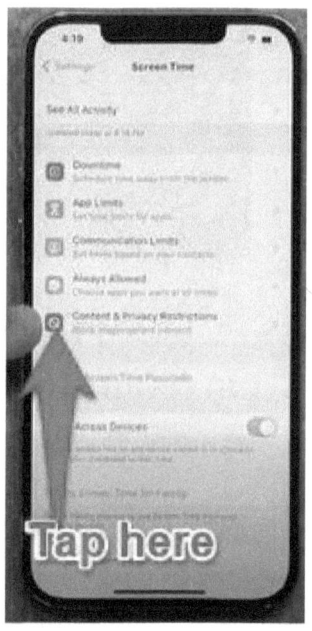

- In the "Content & Privacy Restrictions" Menu, you have three options at the top. You can go into any options to fully customize your phone and restrict access in the areas and apps you want.

- When you click on the first option, "iTunes & App Store Purchases," you will be required to put in your "Screen Time Passcode" that you just set up, and a new menu will pop up.
- In the "iTunes & App Store Purchases" menu, from the list of options available, by tapping on the option to open a new menu with two options: "Allow & Disallow," you can disallow people from installing applications, deleting applications, or making in-app purchases.
- Here is one example, go to the home screen and press and hold one application, i.e., Whatsapp, and it's going to give you a list of options, one of which is the option to "Remove the application."

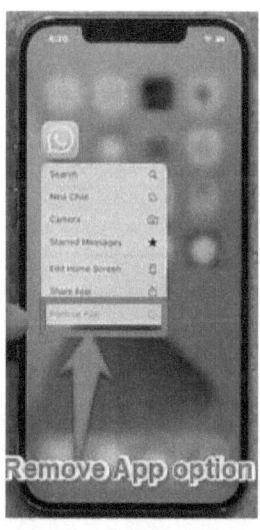

- If you don't want people removing applications when you hand them your phone, even if it is by mistake, you go into Settings, go into Screen Time, go into Content & Privacy Restrictions, go into the iTunes & App Store option, put in your Screen Time Passcode, tap on the "Deleting Apps" option and tap on the option 'Don't Allow."

- When you go back to your home screen and try to press and hold on that application, you cannot delete that application, you won't see see the delete option anywhere.
- When you tap on the "Edit Homescreen" option, there is no "x" symbol on any of the applications on the home screen to delete the appliations.

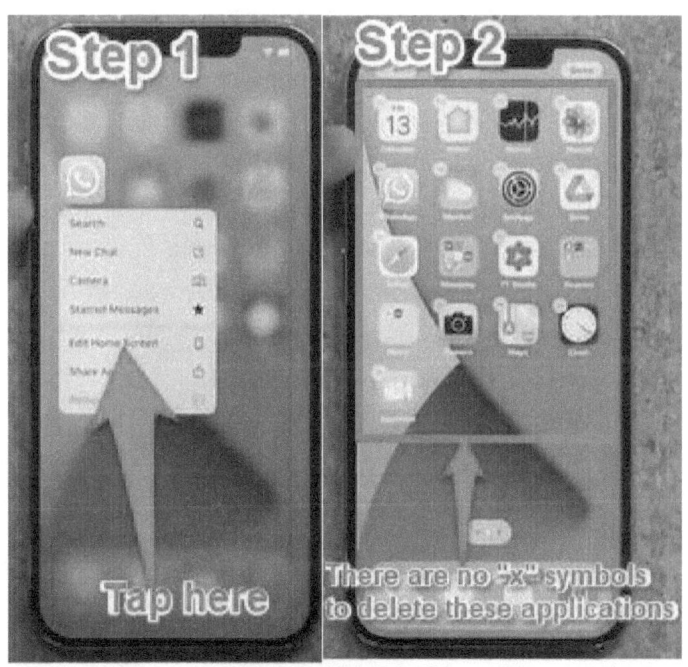

- Back in the "Content & Privacy Restrictions Menu, you can tap on the "Allowed Apps" options, type in your screen time passcode and disable all the applications listed there by tapping on the nubs beside each of them, so people cannot use them unless you allow them.

- So if you disable them, they are going to disapper from the home screen. On the home screen you won't find any of the applications you disabled, in this case, the camera, the mail and so on. So you have this option as well.

- Also, in the Content & Privacy Resrictions Menu, scroll down and you will discover a lot of options that you can take and modify it based on your needs.

Customize Your Control Center

The next thing you have to do has to do with the "Control Center." In order to access the Contol Center, follow these steps:

- Pull down the top right corner of the screen and the Control Center pops up.

- Other than the top portion – the first three rows of the Control Center indicated in the image below, the Control Center is fully customizable at the bottom.

To fully customize the Control Center, follow these steps:

- Go to Settings and tap on the Control Center.
- In the "Control Center" Settings Menu, you will see list out Controls that are included ("Included Controls") and the ones you can also add ("More Controls").
- In the "Included Controls" Features, you can remove the option you want by tapping on the minus icon beside it and an option remove will pop up.
- Click on the remove option and it removes that particular control.

- If you want to use a particular control more than the others, i.e. Alarms, tap and hold on the three parallel line icon beside it and drag it to the position you want it to be on the list, in this case the top.

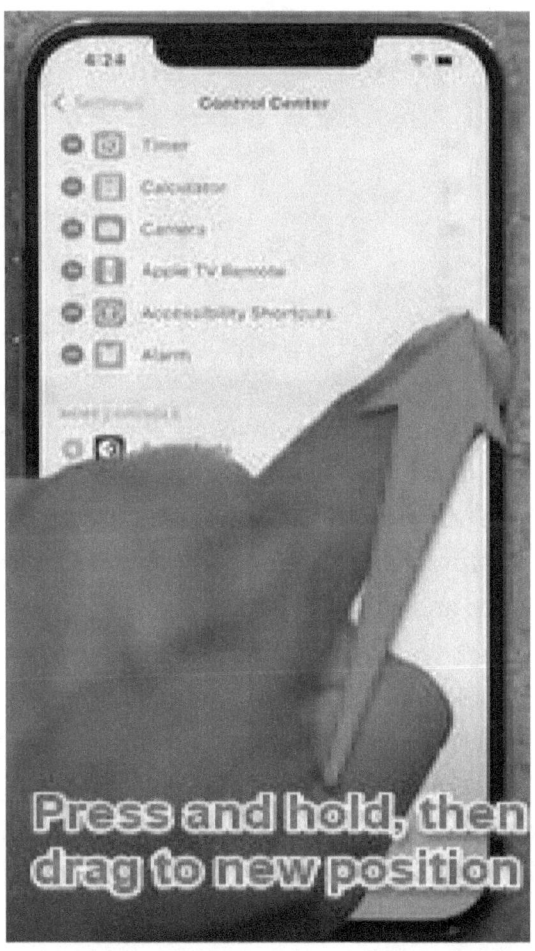

- When you go back to the Control Center, the Alarms icon will be at the top; the first visible icon at the customizable bottom part of the Control Center.

- You can also remove all the controls listed out in the "Included Controls" features and when you go to the Control Center, there are no icons in the customizable bottom part of the Control Center, it is nice and clean.
- Once you have removed all the controls from the "Included Controls" feature, you can add any control you like from the "More Controls" Feature by tapping on the particular control you want to include in the Control Center i.e. Apple TV Remote (if you have an Apple TV, you can use the Apple TV Remote on the device to control the TV), Camera, Calculator and Dark Mode.
- After adding them, go back to the Control Center and you will discover, it is much more customized based on how you want to use it.

Chapter 5

25+ TIPS, TRICKS AND HIDDEN FEATURES FOR iPHONE 12 PRO AND iPHONE 12 PRO MAX

Twenty-five plus (25+) tips and tricks feature for iPhone 12 Pro and iPhone 12 Pro Max. These features will range from essential features to hidden features to massive features so that you can get acquainted with your brand new and shiny iPhone 12 Pro and iPhone 12 Pro Max. These tips are going to enhance your ownership of your smartphone.

The Basics Of Navigation

The first thing to show you has to do with navigation basics, just if you are brand new to the iPhone scene.

- If you are in an application and want to go back to the home screen, swipe the screen from the bottom and back to the home screen.
- If you want to bring up your recent applications, do the same thing, swipe the screen from the bottom to the top and let go, and all your recent applications are going to show up, so you can access anyone you desire.
- If you want to remove any application from the recent applications, swipe it up, and it is gone.
- If you go into an application, you will see a "black bar" at the bottom part of the App screen; use that black bar to switch between applications by grabbing the black bar and sliding over.

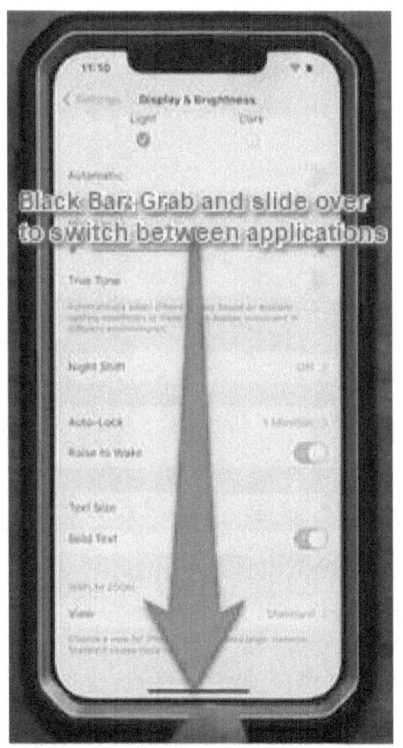

How To Get A Customized Home Button

It is important to note that some people still miss the home button used on iPhones. They can't be able to click on something and go home. You can activate a menu that does exactly that. To do this, follow these steps:
- Tap on Settings and go into "Accessibility."
- In Accessibility, go into "Touch."
- In Touch, go into "AssistiveTouch, once you are in "AssitiveTouch, tap on the nub beside the "AssistiveTouch" feature at the top of the AssistiveTouch page to enable it.

- Once AssistiveTouch is enabled, a round button will appear that floats on the screen. You can put it anywhere on the screen. You can place it at the bottom of the screen directly above the black bar.

- When you tap on it, an options menu will pop up with a "Home" option on it.

- When you click on the "Home" option, it takes you to the Home Screen, and the button remains where you placed it.
- You can access the control center from there on the round button options menu, and you can bring down the notification center.
- You can assess the Home screen with another method. On the "AssitiveTouch" option, click on the option listed as
- "Single Tap," and from the list of options that appears, choose "Home."

- When you click on the round bottom, the single tap always takes you to the Home screen. So it is as if you have an actual button at the bottom of the screen to take you to the Home screen. You don't have to miss your home button again.
- It is also vital; when you go to Settings, you can also configure what you want to do when you double-tap on the round button by going into the "AssistiveTouch" Settings Menu and tapping on the option listed as "Double Tap."
- Once the Double Tap menu appears, select one option from the list of options by tapping on it, in this instance, "App Switcher."
- Now when you double-tap the round button, just like in the good old days, it brings up the app switcher.

- So you can have this software key sitting right where it used to be when we had the physical buttons and behave exactly like in those days.
- You can further customize the round bottom in the "AssistiveTouch" Settings Menu, as much as you want.
- If you want to use the round button as a menu item, tap on the "Single Tap" option in the AssistiveTouch Settings menu and select "Open Menu" among the list of options.
- In this scenario, when you click on the round button, it just expands and gives you additional functions.

How To Back Up Your iPhone To iCloud

- Go to Settings and click on the first option below the "Search" option in the Settings menu called "Apple Cart" (Apple ID); it is where all your information resides in respect to your Apple account.

- Once you are in the Apple cart, tap on the "iCloud" option.
- Once you are in iCloud, scroll down and tap on "iCloud Backup" to ensure your phone is backed up to iCloud by ensuring the iCloud Backup option is enabled.

- So, if something happens to your phone, all your data is backed up, so you can quickly recover it from the iCloud.

How To Get Additional Cloud Storage

By default, everyone gets 5 Gigabytes (GB) of iCloud storage, and you can get additional iCloud storage by purchasing it. To do that, follow these steps:

- On your iCloud Settings in the Apple cart Settings Menu, tap on "Manage Storage."

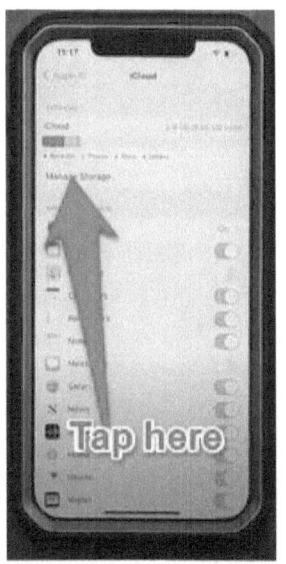

- In "Manage Storage," tap on "Change Storage Plan," and it will give you a list of options or offers through which you can purchase additional iCloud storage.

- After you upgrade your iCloud storage, you can always go back to downgrade by clicking on the "Downgrade Options," if you don't need that storage anymore.

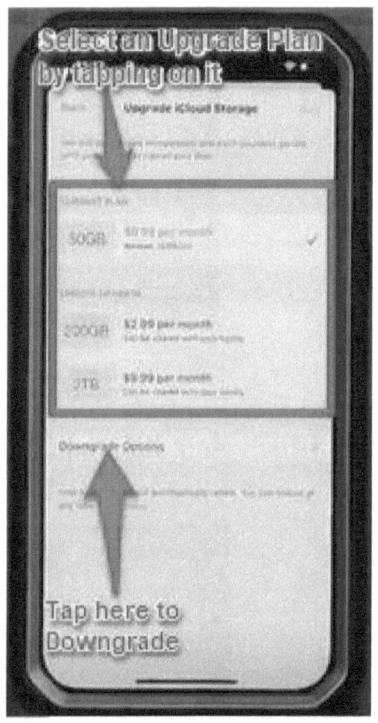

- The additional iCloud storage offers are charged monthly, from month to month to month.

How To Use The Dark Mode Feature In The Brightness Slider

- Go into your Control Center by pulling the screen from the top right corner of the display.

- Press and hold on the brightness slider and it gives you a new menu with the brightness slider and three options.

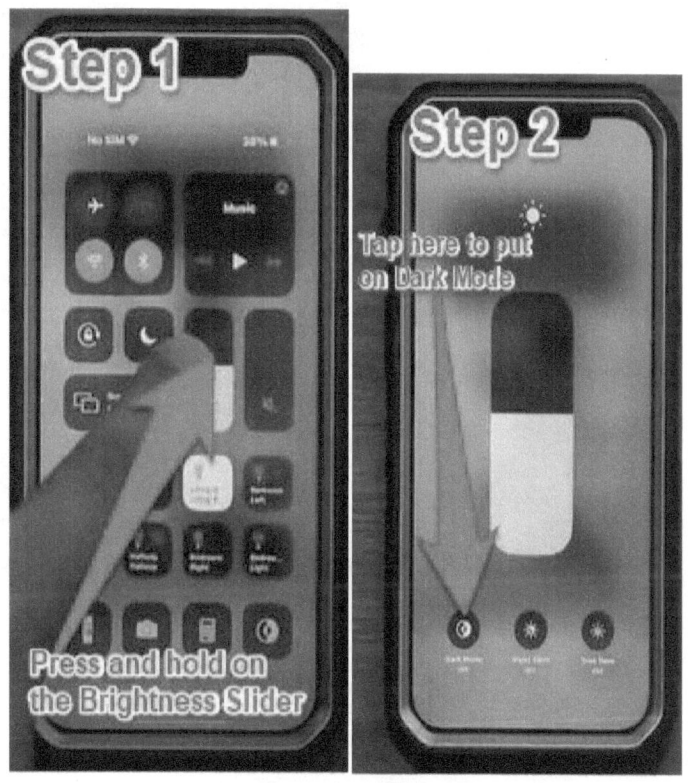

- When you tap on the "Dark Mode" option, even the wallpaper of the device reacts to that option and the wallpaper becomes darker, so it is easy on the eyes and also saves some battery because the display of the device is an oled display.

New Wallpaper Features And How It Increases Battery Life

- Go into your Settings and go into Wallpapers and tap on the "Choose a Wallpaper" option and tap on the "Stills" option.
- Any wallpaper that supports two wallpapers in one, light version and dark version is going to have a little spherical symbol at the bottom and you can choose that wallpaper.

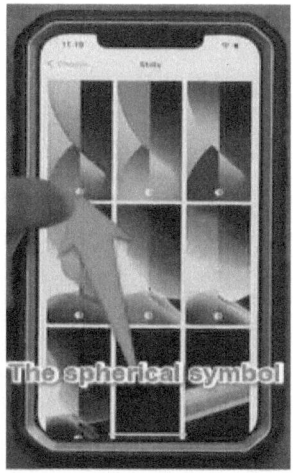

- Otherwise, you can choose anyone of the wallpapers without a symbol at the bottom.
- There is one more thing you need to know, if you choose a black wallpaper {there is a reason they are giving you an "All Black Wallpaper"}, this is what it is going to do, it is going to save you battery.

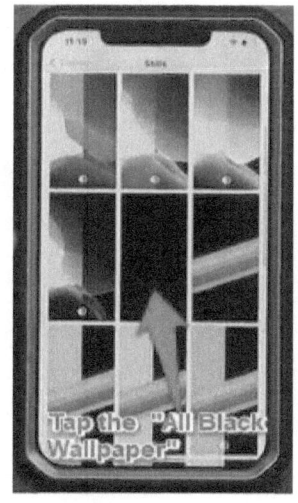

- The device display is an oled display, if the screen is as black as possible, it simply consumes less battery.

- So, if your aim is to increase your battery life, you need to choose the "All Black Wallpaper" because it will certainly increase your device's battery life, or you can go with one of the wallpapaers tht shift from light mode to dark mode based on the choice you make with the dark mode option in the brightness slider.

How To Customize Your Notifications Per Application

- Go to your Settings and go into Notifications.
- In the "Notifications" Settings Menu, you can choose any application you want, i.e. Messages, and modify the way you receive notifications.

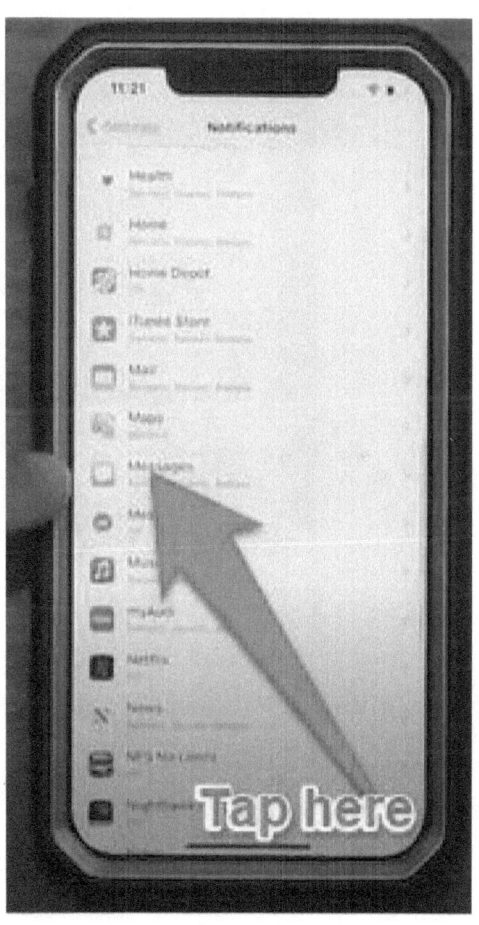

- For example, tap on "Messages,"and a new menu will appear which you can use to modify the way you receive notifications.

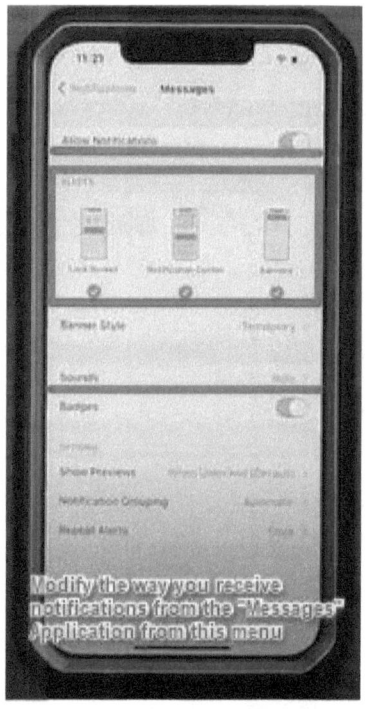

- You can allow or disallow notifications for any application by enabling or disabling the nub beside the "Allow Notifications" option in the applications notification settings menu.
- You can also customize the way you want to get alerted by selecting any of or all the options displayed in the "Alerts" Feature, i.e if you want the alerts to show up on the"Lock Screen,"

"Notification Center," or if you just want a "Banner," a little pop-up banner on the top of your screen.
- You can also customize the sound of each notification for any application by tapping on the "Sound" option and clicking the "Alert Tone" of your choice.
- So you can have a different notification for messages, for imessages and a different notification for WhatsApp or Telegram, if that is what you use.

You can always go to notifications and customize how you get notified for every single application in here and if there is an application that is bothering you too much, you can go there and disable the notification so you never get a notification from that particular application.

How To Take A Screenshot On Your iPhone And How To Use The Screenshot Customization Screen

To take a screenshot on your iPhone, follow these steps:
- Press the "Volume Up Key" and the "Power Key" at the same time.
- Once you do that, it yakes a screenshot and the screenshot which appears as a small image by the bottom left corner of the screen, you can swipe it away and it still gets saved anyway.

- You can also edit the screenshot that appears in a small image. When you take the screenshot, tap on the small image and it is going to expand.

- On the expanded menu, you can select "Pen" to write on the screenshot as you please.
- On the expanded menu, you can tap on "Ruler" and "Pen" to make perfect lines on the screenshot by taping on the ruler and then the pen to draw the perfect lines with the aid of the ruler on the screenshot.

- You can change the size of the screenshot by holding and dragging the bottom right corner of the screenshot to the size you want.
- When you are done editing the screenshot, tap on the icon beside the delete icon at the top right corner of the screenshot and a drop-down menu will appear, with which you can share the screenshot.
- You can also tap on the "Delete" icon by the top right corner of the screen to delete the screenshot, if you are not satisfied with it.

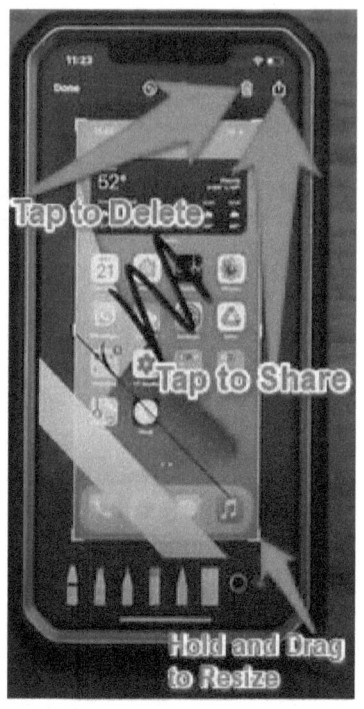

How To Record Your Entire Screen

Another thing you can do with your iPhone 12 Pro and 12 Pro max is, you can record the entire screen and this is very useful if you want to show somebody how to do something or ask a question.

To access the "Screen Recording" option, follow these steps:

- Pull down the control center and you should have a record option in the customizable buttom part of the control center.

- If you don't see the record option, all you do is go to your Settings and go into Control Center.
- Once you are in the Control Center Settings Menu, scroll along the "More Control" Feature till you find the "Screen Recording" option, click on the "Plus" icon beside it.

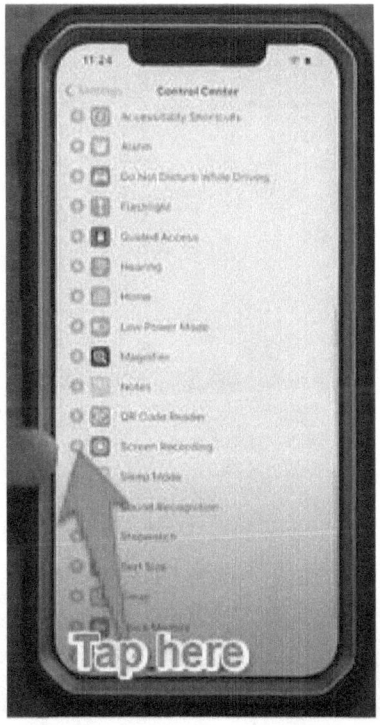

- The "Screen Recording" option will now be included in the "Required Controls" Feature at the top. Which means, it is going to appear in the customizable buttom part of the control center when you pull it down.

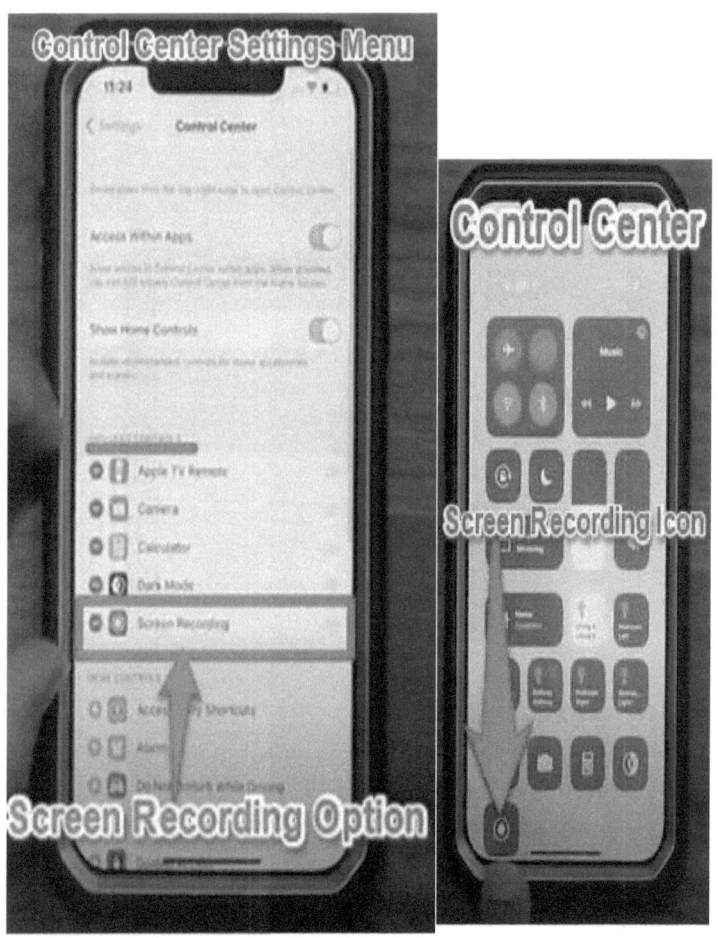

To record your entire screen, follow these steps:
- Bring down the Control Center, press and hold on the "Screen Recording" icon, and an options menu will appear.

- On the Screen Recording options menu, a list of options is displayed. Tap on the option you want to save your screen recording to, i.e., Photos.
- Next, click on the option listed as "Start Recording," and it will countdown from three (3) to one (1).
- After the countdown ends, the device screen is recorded as indicated by the red timer at the screen's top-left corner.
- You can always go back to the Control Center and press and hold on the Screen Recording icon to access its options menu, and on the menu, you can turn "on and off "the microphone by tapping on the "microphone" icon.

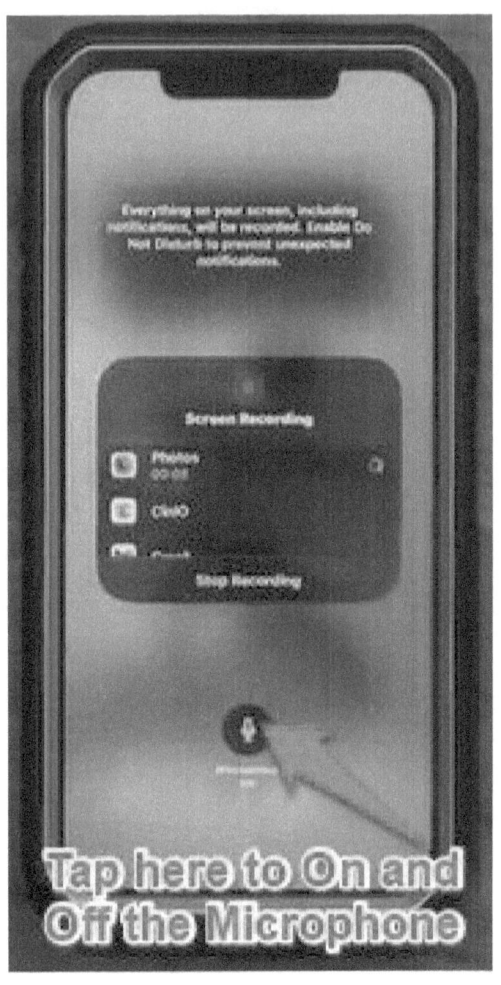

- If you want your voice for instructions, you can add your voice to your recording by putting on the microphone.
- When you are done recording, you tap on the red timer at the top left corner of the device screen, and a message will pop up, asking if you want to stop screen recording. Click on "Stop."

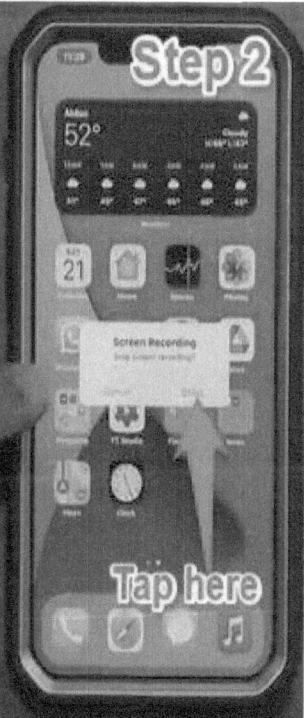

- After clicking on "Stop," another message will pop up at the top of the device screen saying screen recording has been saved to photos (or to whatever App you choose to save the screen recordings.
- You can go into "Photos" and replay the whole screen recording, or you can tap on the message that popped up to replay the whole screen recording.

- You can send the screen recording to anybody, and it is a great way to record your screen built into the phone.

How To Modify The Pixels And Frames Of The Camera In Video Mode Without Leaving The Camera App In Seconds

The Camera is capable of a lot of things, especially the video mode. To modify the "Pixels" and "Frames Per Second" of the Camera in video mode without leaving the camera application, follow these steps:

- Launch the Camera.
- Go into Video Mode, and at the top right corner, you see "HD" and "60." HD means 1080p, and 60 means 60 frames per second. Both of these texts (HD and 60) are clickable.

- If you click on "HD," it switches from HD to 4K.
- If you click the number "60," it switches your frames per second from 60 to 30.
- Now the camera has 4K at 30 frames per second.

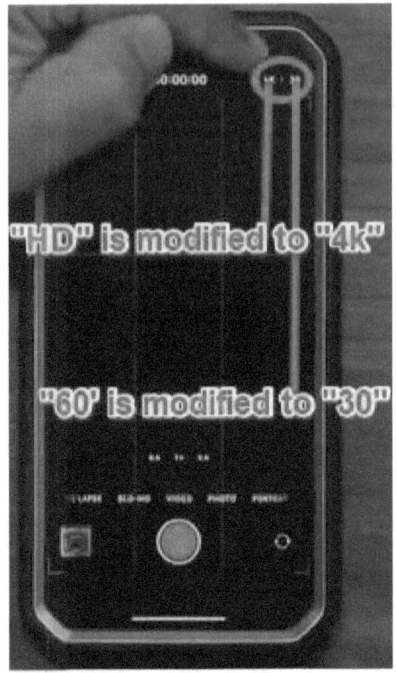

- If you tap on only the "4K" again, you have HD at 30 frames per second.
- If you tap on "30," it switches to "60," and you have HD at 60 frames per second.
- So you don't have to go out of the application; you can modify the Pixels and Frames Per Second right in the camera application.

How To Noties If An Application Is Using Your Device's Microphone Or Cameras In Any Form

You might be wondering why the tiny "Green Light" appears at the top right corner of the screen. When you pull up to the home screen, the light disappears, but you will see the little green light when you bring up the camera.

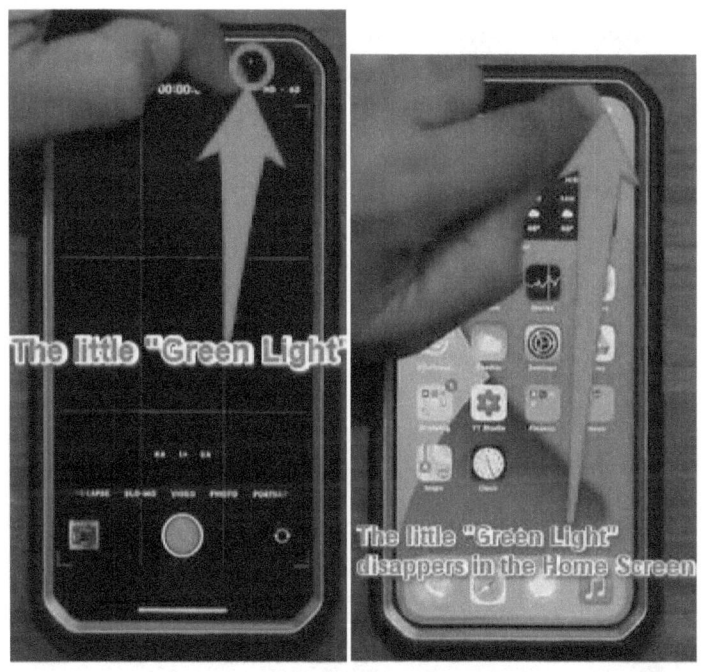

- That "Green Light" means both the microphones and the cameras are currently in use.
- The "Green Light" is going to show up in any application that is using your device's microphone or cameras in any form, just so you are aware that you are being recorded somehow.

How To Access The Different Formats In The Camera Settings

- Tap on the "Settings," scroll down till you find the "Camera" option. Tap on it.
- In the Camera Settings Menu, tap on the option 'Record Video," and a new menu appears from where you can access the Formats Setting to enable you to switch between different camera settings.

How To Enable The High Dynamic Range (HDR) And Dolby Vision

The iPhone 12 Pro and 12 Pro Max allows you to record in High Dynamic Range (HDR) and Dolby Vision.

- To enable it, tap on the nub beside the "HDR Vision (High Efficiency)" option listed in the Record Video menu in the Camera Setting Menu, and you can record in HDR and Dolby Vision.

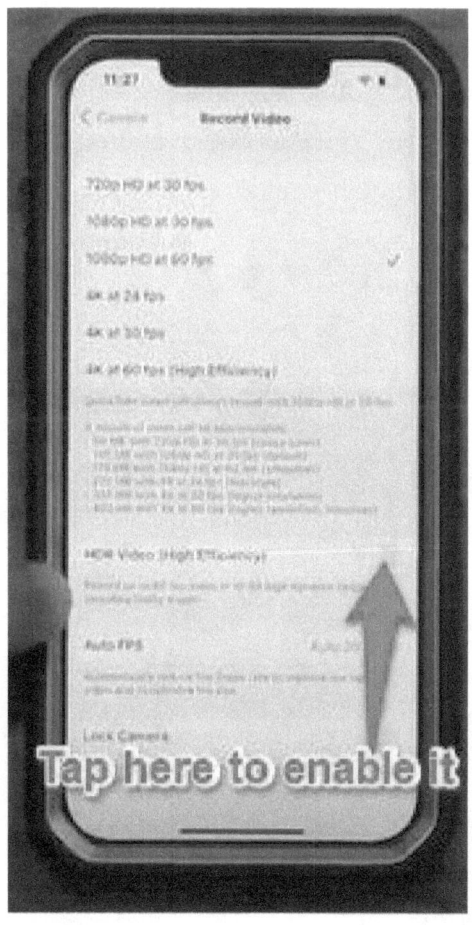

You should disable the "HDR Vision (High Efficiency)" option unless you are specifically interested in HDR with Dolby Vision. Otherwise, stick to the different camera settings options and pick one of those options for your recording, and you are going to be good to go.

How To Access The App Library

All the device apps are now in the "App Library," and the App Library is going to be around the corner. So you can customize your screens however you want, and then you swipe over and get your App Library.

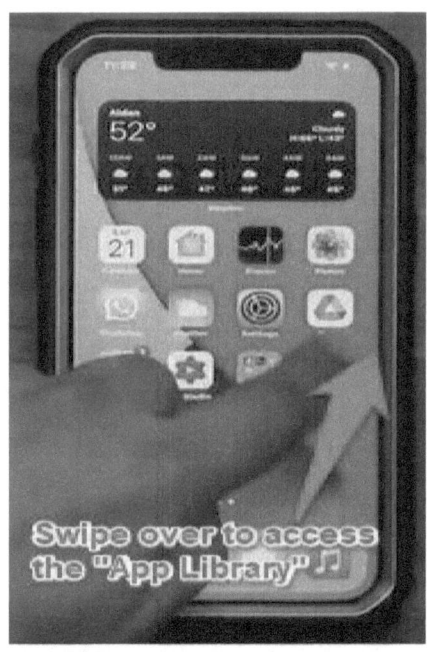

- In the App Library, you can search all your applications from there by tapping on the "Search"

icon and typing the name of the application you want to access.

- In the "Search" Menu, you can also choose the application by letter by tapping on the specific letter you want on the alphabet list by the right side of the device. So when you tap on the letter "r," all the apps starting will the letter "r" will appear.
- You can also swipe the apps up and down to get the specific app you want.

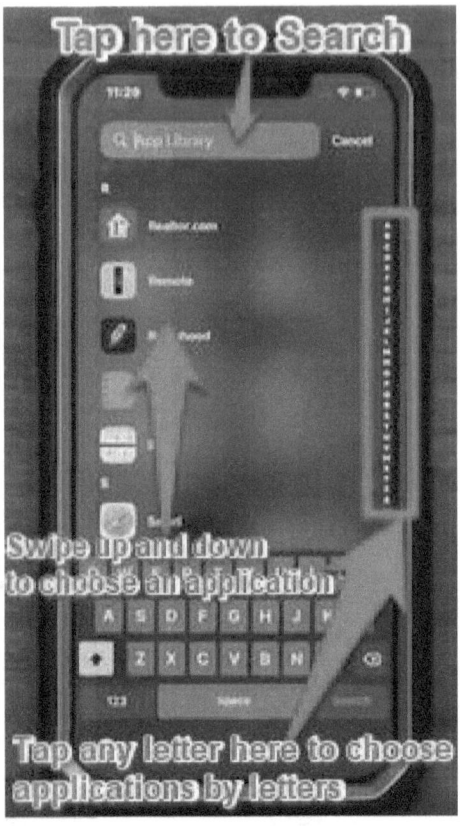

Looking at the iPhone Pro Max is very big on hand and has a larger screen. To access the control center by swiping down the device screen once, you will need to customize the device.

To use the one hand gesture, follow the step below:

How to Use Reachability feature

- To enable this feature, tap on "Settings" and scroll down, then tap on "Accessibility."
- A drop-down menu will display with a list of features, scroll down and tap on "Touch."

- Another menu will pop up with a list of features; tap on the second feature, "Reachability."
- If the "Reachability" is turn on, go back to the device's home screen, and tap on the "Home button "it will pop up Reachability. From the Reachability point, you will swipe down with one finger and launch the control center.

Smart HDR Three

The Smart HDR Three is a new feature of the iPhone 12 Pro Max. It is a unique setting that you should apply or enables In the device setting to maximize the device for optimum performance on the live camera and video App. To enable this feature, tap on "Settings," then scroll down and tap on "Camera" in here Smart HDR feature will be displayed.

How To Use Back Tap

Another brand new feature of the iPhone 12 Pro Max is the "Back Tap," the back tap feature is a unique gesture that you should enable for optimum navigation. Follow this step below:

- If you tap on the back of the iPhone 12 Pro Max twice or three times, the control center will pop up even if there is a back cover on the device. Triple tap again for the second time the control panel or center will disappear
- To enable this feature, tap on "Settings" and scroll down, then tap on "Accessibility."
- A drop-down menu will display with a list of features, scroll down and tap on "Touch."
- Then another menu will pop up. Scroll down and tap on "Back Tap."
- This will pop up two lists of an item, the "Double Tap and Tripple Tap."
- Tap on any preferred one, for me; my best bet is the "Double Tap." but for the triple tap, when you tap

on it, so many gestures will pop up that you can use to navigate through the device

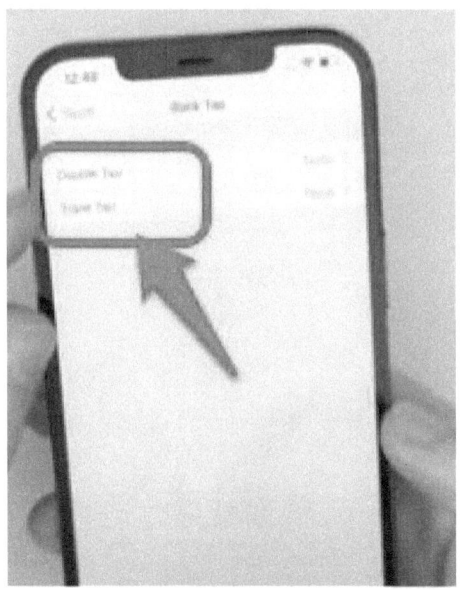

How To Hide Page

On the iPhone 12 Pro Max, you can always hide the device's screen page; so if you have many pages on the screen. to "Hide Page," follow this step below:
- Tap and hold on anywhere on the screen to juggle mode. You will see bubble around the page "Dots" at the bottom part of the screen device
- Click on the "Dot," because it is clickable
- **Note**: an interface will pop up on the screen, and you can hide a full page in there
- When you tap and hold down the home screen, the hidden page will display with an icon button at the bottom of the page

- Please tap on the button icon under the page screen to enable hide page on the springboard
- To unhide the page, tap on the same button

Chapter 6

BEST ACCESSORIES FOR YOUR iPHONE 12 PRO & iPHONE 12 PRO MAX

The accessories listed in this chapter are compatible with the iPhone 12 series. They can make the device a bit safer, a bit easier to use, and get an overall better experience with your device.

MagSafe Accessories

- Apple Silicon Case
- Apple 20W Charging Brick
- 10,000 milliamp hour Anker external battery with USB-A and USB-C
- Yootech compatible with MagSafe

Other Accessories For The iPhone 12 Series

- Air Pods Pro
- Apple Watch Series 3, SE and Series 6
- Ulanzi ST-09 Apple Watch Mount
- Ailum Glass Screen Guard
- OtherBox Defender Case
- Anker Nano iPhone Charger, 20W PIQ 2.0
- Aukey Omnia 20W Charger Fast
- Nomad Base Station Pro Wireless Charger
- Mighty Vibe with Bluetooth
- Updated Clutch V2 Battery (External Battery)
- inCharge 6 Cables
- Manfrotto Clamps and Tripod
- Flydigi Shadow Stinger Triggers
- Razer Kishi Controller

Apple Silicon Case

The cool thing about this latest Apple Silicone Case is the ring in the front of the case is the MagSafe Magnetic part of the case, and they come with it inbuilt. It is effortless to use. Put the iPhone in the case. One thing about the iPhone 12 Pro is its stainless steel sides. Apple designs them with the most 'finger printable' material. The sides of the device easily pick up fingerprints, same with the Apple Silicone Case.

MagSafe is not only for charging; it does some interesting things with these Apple Silicon cases. So, Apple is using a magnetometer and NFC to recognize the devices that are connected to it.

Lock the display and then put it on the Silicon case. You will discover that the MagSafe shows up as the case's color when it connects to the case. It is using that NFC in the back to do that. If you take the case off and turn off the display, put on another colored Silicon case. The MagSafe connects, and it shows you the color of the new case, so it knows that it is connected to this new case.

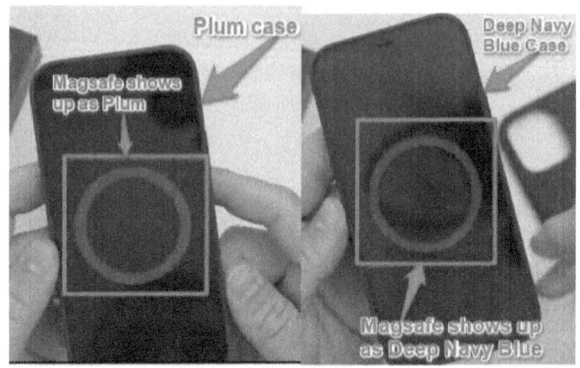

Of course, you can charge through it, and the magnets hold on tight, especially with the case, because the case is doubling the magnet's power, so it is really on there. You can easily pop off the MagSafe to remove it.

These cases are fantastic because they protect the bottom of the device this time too.

Ailum Glass Screen Guard

After getting the Ailum Glass Screen Protector, installing it on the iPhone 12 or 12 Pro Max is effortless. First, clean the screen with the wet wipes, and next, dry the screen with the Dry wipes. If there is still some dust on the screen, remove it with the Dust-absorber. Peel off the protective film on the Ailun Screen Guard carefully. Align the screen guard cutouts and edges with the phone screen correctly. Then tap the center of the screen protector and wait for a few seconds to let the screen protector stick on the screen automatically.

It has a "Privacy Feature" that people cannot see your screen from the other side than regular tempered glass when you install it. You can cut paper with a paper cutter on the device screen with the screen guard installed, and it won't leave a scratch. If a drop of waterfalls on the screen protected by the screen guard, it doesn't spread on the device, and the water drop falls up, when you write with a marjer too it isn't visible on the screen, you can clean the screen to clear off the traces of water or ink left behind.

Nomad Base Station Pro

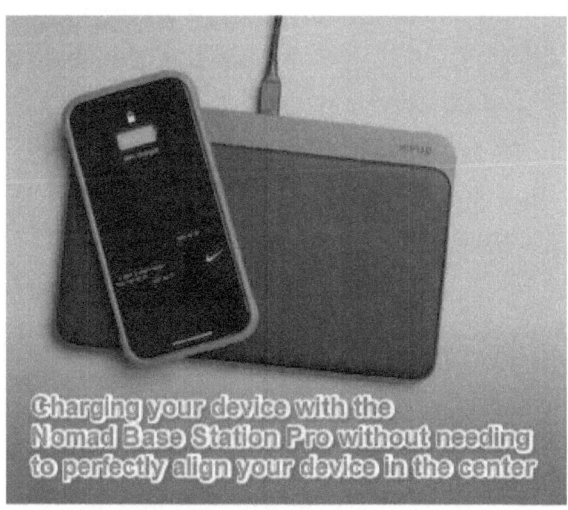

Charging your device with the Nomad Base Station Pro without needing to perfectly align your device in the center

It is the most expensive wireless charger you may ever own but it will probably be the last one you ever need.

First of, it looks super sleek and with Nomad does with all there gear, it's got a soft leather padded top layer stuck into a precision machine aluminium frame. The only thing it does is wirelessly charge your devices.

This charger doesn't require you to correctly align your device in the center for it to start charging, and that's the convenience you are paying for. This is the first of its kind.
It can also detect up to three (3) devices without having to align them in a specific way, and Apple's MagSafe charger also solves that issue, but with MagSafe, you are stuck with one charging device at a time.

Charging Brick

One of the things Apple did in 2020 was they are taking a more "Carbon neutral stance," and to do that, they have shrunk the size of their packaging, and they did not include a charging brick with the iPhone 12 series. They gave us the cable but not the charging brick. So you are going to need a charging adaptor if you don't have one.

The cable that comes with the iPhone 12 series is the USB C, not USB A to lightning cable, so a regular brick will not work with the new cable unless you already have an old brick and an old cable. The Apple 20W charger is recommended. It is more powerful than the older Apple chargers. It charges a little bit faster than the standard Apple charging brick that you might you. It would give you the best power delivery that the MagSafe charger can do when it works with MagSafe. This charger is also a more future-proof model, a little more expensive, but it will work with all of Apple's things going forward. It already has USB C, and it does a little faster charging if you are in a hurry.

Manfrotto Clamps And Tripods

One powerful thing about the iPhones is the camera accessories; to take full advantage of the camera is essential. Manfrotto can hold the iPhone 12 Pro and Max for effective shooting like little phone clamps. All you do is clip it onto the phone, and then it gives you a few quarter 20 threads so that you can adapt it to a tripod (A Manfrotto mini tripod) if you want to.

You can use the Manfrotto Clamps with Manfrotto tripods for zoom meetings and all kinds of things. Even if you don't have a tripod, this version of the Manfrotto clamp has a "little kickstand" on the back. You could use it to watch movies or do a zoom meeting without having to hold the device off of just a standard desk. The little kickstand folds back onto the back of itself. The build quality is pretty good, and it gives you that extra advantage with the camera.

Chapter 7

WHICH APPLE WATCH IS RIGHT FOR YOU? IS IT THE SERIES 3, OR THE SE OR THE SERIES 6?

To help you decide which Apple Watch is right for you, I will compare the three available Apple Watch if you were to go to an Apple Store. They are the three options you can buy right now as far as the series of Apple watches. You may find others available elsewhere, but these three are the only ones currently sold new from apple.

First off, the Series 3 Apple Watch in 42-millimeter size, Apple Watch SE in 40-millimeter size, the newer SE, and lastly, the Apple Watch Series 6 in 44 millimeters and product red, one of the new colors.

Prices And How They Compare With One Another

Apple Watch Series 3

For the Series 3 Apple Watch, Apple only offers it in a couple of different variants. So far, they offer an Aluminum with Space Gray and Silver, that's as far as the colors, and then, they also offer it for One Hundred and Ninety-Nine Dollars ($199) for 38 millimeters or Two Hundred and Twenty-Nine Dollars ($229) for 42 millimeters. They offer it with the Ion-X Glass, our case study, and it does scratch relatively quickly. It is cheaper, and it does not come in the other variants. When it was new, Apple offered such as stainless steel or any of the others. You can only get it in Aluminium now. You can only get it with Wi-Fi. There is no cellular available for this particular model.

Apple Watch Se

The Apple Watch SE is available in different variants, for example, Two Hundred and Seventy-Nine Dollars ($279) for the 40 millimeter Apple Watch SE and Three Hundred and Nine Dollars ($309) for the 44 millimeters Apple Watch SE. If you want to add Cellular, you will need to add Dollars ($30). There is a new family setup within the iPhone app That you can allow your child to have, an Apple Watch paired to your phone directly. And so if you check the iPhone, you will see your watches paired to your phone directly, and so if you go into "My Watches," by tapping on the option by the upper left side of the phone, you will see the three watches that are currently paired, and you can add other watches. Still, you will need a cellular watch to add it. So, that means your option is the SE or the Series 6, and that is what you can get to add your child.

You can also use the SE on a monthly plan to set up different things. Now, you can't do that with a Series 3.

The SE is available in Aluminum, and it is also available in Gold, Silver, and Space Gray. This case study is the Space Gray with the red product brand, but you can change out any band you want with it. It is a very similar watch to what we have with the Series 6.

Apple Watch Series 6

With the Series 6 Apple Watch, you have got a couple of different options. Since this is the latest, we have Three Hundred and Ninety-Nine Dollars ($399) for 40 millimeters. Apple Watch Series 6 and Four Hundred and Twenty-Nine Thousand Dollars ($429) for the 44 millimeters Apple Watch Series 6. Again, if you want to add GPS and Cellular, You have to add Thirty Dollars ($30), plus, you will have to have a monthly plan.

The thing with the Series 6 is from our case study, it is in Red, and it also comes in Blue, Gold, Silver, Space Gray and Aluminum, Stainless Steel, and Titanium. You can go all the way up to One Thousand, Four Hundred and Ninety-Nine Dollars ($1499) for a 44 millimeter Hermes Model and so you can go from our case study edition to the titanium edition.

That's the difference between the watches themselves as far as the options. Still, there are a lot of options with Series 6. For example, suppose you get the Aluminium. In that case, you have got Ion-X Glass (as all Aluminium models come with Ion-X Glass), if you get the Stainless Steel or Titanium, you have got the Sapphire Crystal which doesn't scratch very easily. It is worthy of note that it is not a super high-quality sapphire. With the Ion-X Glass, it will get a scratch on it from time to time if you don't use a screen protector. That's something to keep in mind. They also have other colors, such as Space Black and Graphite, depending on if you have Titanium or Stainless Steel, so there are all of those editions. On top of that, there is a Nike edition, which has a couple of different watch faces. So, you have all of those different versions in terms of size and price, so you can start at One Hundred and Ninety-Nine Dollars ($199) and go all the way up to One Thusand, Four Hundred and Forty-Four Dollars ($1499).

Displays For All Three Watchs

Now when it comes to the Displays, this is a huge differentiating factor between all three watches.

For Series 3, it is Edge to Edge and squared off. On the SE and Series 6, the display is rounded and slightly bigger on the 44 millimeters, and you have that huge difference.

Also, SE and the Series 6 displays are a little bit brighter, so they are easier to see out in the sunlight, and also they share the same display, but the Series 6 has an "Always-On Display." For example, when Series 6 is worn on your wrist and tilt your wrist off the side, the display goes to sleep after a moment, but it is still on. It is always on. So you have that option with the Series 6. You do not get that option with the SE or Series 3.

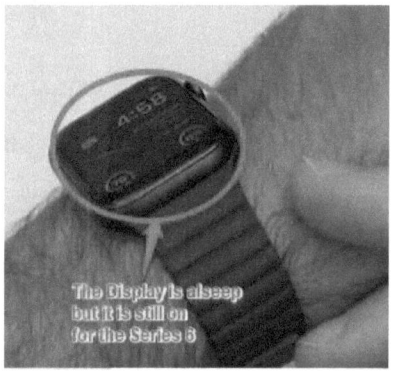

Chapter 8

SIRI

In this chapter, you will learn more about using "Siri" to send messages and some other items. For those who know how to use it, I will show you something new on Siri.

To learn more about "Siri," follow this Step below

Siri

- On your iPhone screen, tap on the "Settings," a drop-down menu will appear with different settings.
- Scroll down and tap on the "Siri and Search," then turn off the Listen for "Hey Siri" then, after these settings, you turn it on later.
- Pressing and holding on the side key button will activate the "Siri," you can also do it on the lock screen, but it has some limitations if you go through the lock screen. So it is better to use the side key; when the side key button is held down, Siri automatically pop-up at the bottom of the screen. Please note that if you only press on the power key button, it will lock the device, so make sure you press and hold on the power key button

- **Allow Siri**: When locked," this can be achieved when your iPhone 12 Pro device is on a lock screen. If you turn the "Allow Siri when locked on." For me, please turn it off
- **Language**: When you tap on "Language," a drop-down menu will pop-up, in here you will see so many languages that you can select for Siri to communicate with; it all depends on you.
- **Siri voice:** The said default voice of the Siri is an American female voice. But you can change this female voice to any other voice. When you tap on the "Siri voice," immediately it will display another drop-down menu, in here; you will see so many Siri voice that you can always change to for use. When you tap on any of the Siri's voice immediately, it will start downloading. When the download is complete, you can use it.
- **Siri responses**: With "Siri responses" settings, Siri can always respond even when the smartphone is on silent mode. To activate this, tap on the "Siri responses," and this will pop-up with a list of settings. Then "Click" on the "Always." You can as well tap on the "When silent mode is off." You can also use the third command Only when "Hey Siri," Siri can only
- Respond to you if you use the word "Hey Siri" only when it is set as the default, so any other word use in place of the Hey Siri will be a waste of time.

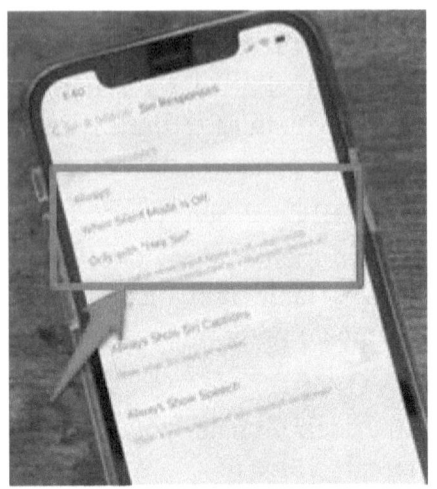

- **Always Show Siri Captions**: It is essential to know about this setting of the Siri before turning it on. The "Always show caption," when activated it will write out what Siri we say on the device screen in a small box like a square, and it enables you to understand the words of Siri clearly

- **Always Show Speech**: This tool is similar to "Always show Siri Captions," it also show the transcript of what you say to Siri on the device screen when it is activated. This give you a better understand of what you say to Siri if it is properly spoken or not by displaying at the bottom part of the screen. So make the both settings are turn on
- "Announce Massages with Siri" this setting is unique, when Announce Masages is activated or turn on Siri can read out any massages that comes into the device with else. With the second generation Airpod you can respond back to the massages immediately without saying the key word "Hey Siri" if it is connected with know interruption

Create Your Own Contact With Siri

Siri is a unique tool and a problem solver for the iPhone 12 pro and max users; with Siri, you can create your information base on your contact list. Follow this step below to activate it:
- Tap on the settings on the device screen
- A drop-down menu will display, scroll down and tap on "Siri and Search," another menu will appear.

- Tap on "My Information," and your contact list will open up.
- Then create your contact list, and this will have all your information such as your id number, your phone numbers, Email address, and Residential Address e.t.c
- When this information is created, Siri knows how to act on it. For example, with this information, when you say Siri take me home or remind me to visit the supper market by 9.30 am base on the information you gave, Siri will automatically remind you.

Chapter 9

TIPS AND TRICK OF THE CAMERA OF THE iPHONE 12 PRO MAX

If you want to use the iPhone 12 Pro Max camera for efficient and effective performance and to ensure that your photography is at the next level with less effort, follow this step below:

How To Enable High Efficiency

- Tap on the "Settings" of the device
- Scroll down and tap on "Camera" a drop-down menu will display; in here, tap on "Formats."
- When the format is tap on, two settings options will display (High Efficiency and Most Compatible).
- **Please note,** tap on the "High Efficiency." When the High efficiency is tap on it will automatically reduce the file size and also allow the (4K at 60K frame per second by 1080p at 240 frames per second Slo-Mo and HDR)

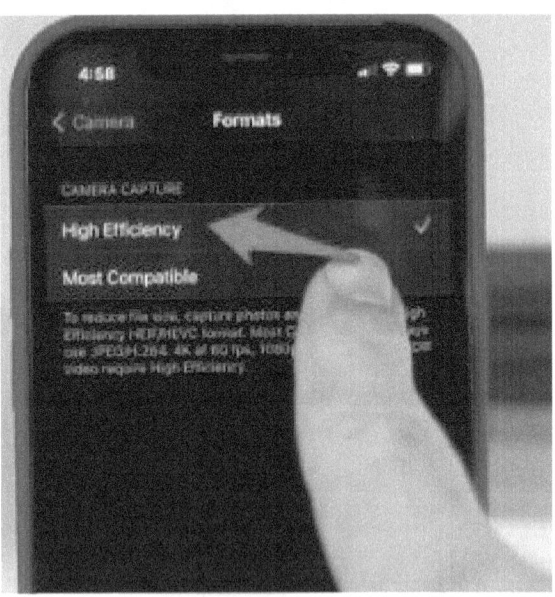

- **Note** when you tap on the "Most Compatible," it will change the format for the video and photo to Jpeg and H.264, and you can be able to use (4K at 60K frame per second 1080 at 240 frames per second. That means the "Most compatible" can't reduce the file size

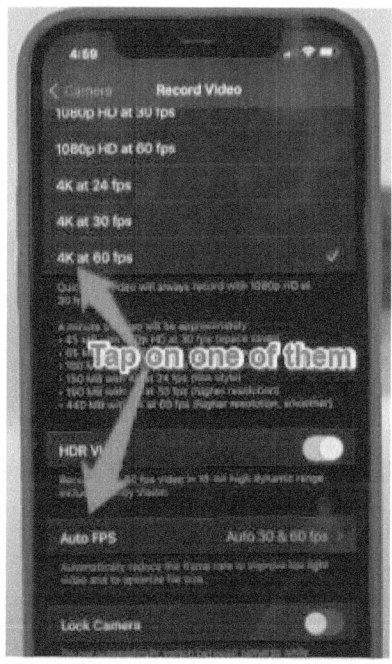

How To Enable Record Video

In the record video settings, you need to set up many advanced features to maximize quality. To enhance the video quality follow this step below:
- Tap on the "Settings" of the device
- Scroll down and tap on "Camera" a drop-down menu will appear
- Then tap on "Record Video."

- Scroll down and tap on this option "4K at 60K fps" so you get the more robust video for high quality with the highest refresh rate

- Also, scroll down and enable the "Auto FPS" this will automatically reduce the frame rate to improve the low light quality of the video to 25fps, and this will be better because it will make have a quick capture of any fast movement, gaming, football games on the pitch and where there is a lot of movement

How To Enable Lock Camera

The "lock Camera" settings are a unique feature to use most, especially when using the standard Lens Camera. When you set it up, it allows the Camera to zoom in three times while recording. When recording, the Camera also captures the object precisely the way it is when you zoom in, and it will avoid jumping.

When it is the regular Camera, it usually jumps to either the ultra-wide or the wide Camera while recording. So if you enable the Lock Camera when recording video at the same time zooming, it won't jump. When using the wide Camera too for recording with the "Lock Camera" feature, you can zoom up to six times, and the Camera will not also jump while the object remains the same.

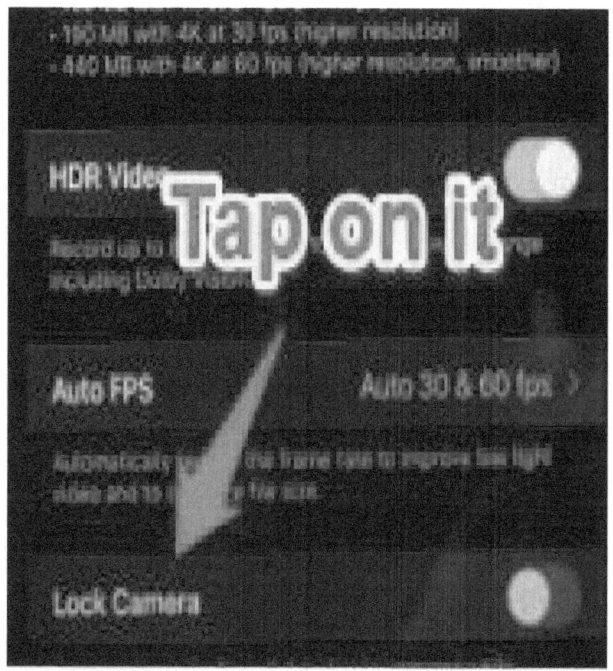

How To Turn On Record Slo-Mo

The Slo-Mo feature is another unique feature that you need to turn on to have a good quality live Camera and photo. Now to check it on, follow this step below:
- Tap on the "Settings" of the device
- Scroll down and tap on "Camera" a drop-down menu will appear
- Tap on "Record Slo-Mo," and it will display two lists of features
- Then pick on the "1080-240fps, and it will always give the very quality you want to get on the video and photo.

How To Use Record Stereo Sound

The "Record Stereo Sound" has many features that keep specific settings inside the Camera App. You should "Check On" only two of these features (Live photo and Exposure Adjustment). When the "Live Photo" and the Exposure Adjustment is turn on, they will automatically apply to the Camera App.

If you use the camera for recording and leave the camera screen and again tap on the camera, it will display the same way you left it before.

How To Use Grid

The Camera Grid of the iPhone 12 Pro and iPhone 12 Pro Max is as well essential to use because it places a frame to the object that you want to take a shot of and centralizes it for perfect optimization

Mirror Front Camera And View Outside The Frame

The "Mirror Front camera" and "View Outside the Frame" is what you should set on because it mirrored the photo when you take a shot. And this makes the image come out in a mirrored quality automatically instead of editing the image manually. So make sure you check it on because it will save you time for editing.

Three Ways To Clear And Free iPhone Ram Memory On iOS 14

Step 1

- If you use your iPhone for a complex activity, I bet you. It would be best to have more of a bigger space, Ram, on the device memory. To clear up space on the RAM of iOS 14, follow the step below:

- Tap on "Settings"
- Tap on Accessibility"
- Then tap on "Touch."
- Scroll down, then tap on "Assistivetouch" to enable it
- Go to the volume up and down key button of the device and the side key button, then tap and hold both.
- It will take you to power off-screen.
- At the bottom right corner of the power off menu screen, tap on the icon of the "Assisted Touch button."
- Again, tap and hold to the home button, it will automatically take you to the device's home screen, and this is where you have clear or free the Ram Memory of the iOS 14. This method required the system touch of which can free the memory of the device.

Step two

To use the second method, you need a free app. The "App" can be downloaded from the "App Store." This free App is called "Cpux," and it gives you all the details you need to know regarding the device. To use the "Cpux App," follow the step below:
- On the iPhone device, go to the "Apple store" and download the "Cpux App" then installed it on the device.
- Tap on the "Cpux App" to open it
- A drop-down list will display
- Tap on "Memory," another menu list will appear.
- Take your eyes to "Capacity," under the capacity; you will see this item (Free, Active, Inactive and Wire, etc.), then check the free space left under "free."
- For instance, the space left on the device is 180, and you want to free up the "RAM."
- Tap on the "Free Up" immediately; the device's RAM will start optimizing its memory, and when it finishes optimizing, it will free up to 500megabytes on the device. So with a couple of clicks on the free up, you will get an excellent free space.

Step Three

This last method to free the Ram Memory is straightforward if you follow this step below:
- On the device's home screen, tap on the "Camera App" to launch the Camera.
- Once the Camera App is launch, close it, go back to the home screen, tap on the "recent icon" button, and swipe up all the recent open app to close it

completely. And automatically, you will free enough RAM space on the device.
- With the method, you do not need a third-party app or trick to free up the device's memory. Do it exactly how it is explained, and it will free up to 500megabytes of the Ram Memory.

About the Author

David Great is a tech writer and owns a blog where he writes about latest tech news and innovations. He is a geek and passionately follows latest technical and technological trends. David holds a Bachelor's degree in Information Communication Technology from New York State University.

He lives in New York with his wife Tabitha and two beautiful children.

www.ingramcontent.com/pod-product-compliance
Lightning Source LLC
Chambersburg PA
CBHW031418210526
45464CB00005B/1951